T0184656

Wissenschaftliche Reihe Fahrzeugtechnik Universität Stuttgart

Reihe herausgegeben von
M. Bargende, Stuttgart, Deutschland
H.-C. Reuss, Stuttgart, Deutschland
J. Wiedemann, Stuttgart, Deutschland

Das Institut für Verbrennungsmotoren und Kraftfahrwesen (IVK) an der Universität Stuttgart erforscht, entwickelt, appliziert und erprobt, in enger Zusammenarbeit mit der Industrie, Elemente bzw. Technologien aus dem Bereich moderner Fahrzeugkonzepte. Das Institut gliedert sich in die drei Bereiche Kraftfahrwesen, Fahrzeugantriebe und Kraftfahrzeug-Mechatronik. Aufgabe dieser Bereiche ist die Ausarbeitung des Themengebietes im Prüfstandsbetrieb, in Theorie und Simulation. Schwerpunkte des Kraftfahrwesens sind hierbei die Aerodynamik, Akustik (NVH), Fahrdynamik und Fahrermodellierung, Leichtbau, Sicherheit, Kraftübertragung sowie Energie und Thermomanagement – auch in Verbindung mit hybriden und batterieelektrischen Fahrzeugkonzepten. Der Bereich Fahrzeugantriebe widmet sich den Themen Brennverfahrensentwicklung einschließlich Regelungs- und Steuerungskonzeptionen bei zugleich minimierten Emissionen, komplexe Abgasnachbehandlung, Aufladesysteme und -strategien, Hybridsysteme und Betriebsstrategien sowie mechanisch-akustischen Fragestellungen. Themen der Kraftfahrzeug-Mechatronik sind die Antriebsstrangregelung/Hybride, Elektromobilität, Bordnetz und Energiemanagement, Funktions- und Softwareentwicklung sowie Test und Diagnose. Die Erfüllung dieser Aufgaben wird prüfstandsseitig neben vielem anderen unterstützt durch 19 Motorenprüfstände, zwei Rollenprüfstände, einen 1:1-Fahrsimulator, einen Antriebsstrangprüfstand, einen Thermowindkanal sowie einen 1:1-Aeroakustikwindkanal. Die wissenschaftliche Reihe „Fahrzeugtechnik Universität Stuttgart" präsentiert über die am Institut entstandenen Promotionen die hervorragenden Arbeitsergebnisse der Forschungstätigkeiten am IVK.

Reihe herausgegeben von

Prof. Dr.-Ing. Michael Bargende
Lehrstuhl Fahrzeugantriebe
Institut für Verbrennungsmotoren und
Kraftfahrwesen, Universität Stuttgart
Stuttgart, Deutschland

Prof. Dr.-Ing. Jochen Wiedemann
Lehrstuhl Kraftfahrwesen
Institut für Verbrennungsmotoren und
Kraftfahrwesen, Universität Stuttgart
Stuttgart, Deutschland

Prof. Dr.-Ing. Hans-Christian Reuss
Lehrstuhl Kraftfahrzeugmechatronik
Institut für Verbrennungsmotoren und
Kraftfahrwesen, Universität Stuttgart
Stuttgart, Deutschland

Weitere Bände in der Reihe http://www.springer.com/series/13535

Alexander Fandakov

A Phenomenological Knock Model for the Development of Future Engine Concepts

Springer Vieweg

Alexander Fandakov
Institute of Internal Combustion Engines
and Automotive Engineering (IVK)
University of Stuttgart
Stuttgart, Germany

Zugl.: Dissertation, University of Stuttgart, 2018

D93

ISSN 2567-0042   ISSN 2567-0352   (electronic)
Wissenschaftliche Reihe Fahrzeugtechnik Universität Stuttgart
ISBN 978-3-658-24874-1   ISBN 978-3-658-24875-8   (eBook)
https://doi.org/10.1007/978-3-658-24875-8

Library of Congress Control Number: 2018965240

Springer Vieweg
© Springer Fachmedien Wiesbaden GmbH, part of Springer Nature 2019

This Springer Vieweg imprint is published by the registered company Springer Fachmedien Wiesbaden GmbH part of Springer Nature
The registered company address is: Abraham-Lincoln-Str. 46, 65189 Wiesbaden, Germany

Acknowledgements

The model presented in this thesis was developed in the course of my work at the Institute of Internal Combustion Engines and Automotive Engineering (IVK) of the University of Stuttgart.

I would first like to acknowledge my gratitude to Prof. Dr.-Ing. Michael Bargende for his outstanding support, guidance and the numerous motivating discussions.

Next, I would like to thank Prof. Dipl.-Ing. Dr. techn. Helmut Eichlseder for his interest and for joining the doctoral committee.

This work would not have been the same without the valuable support from various people at the Institute of Internal Combustion Engines and Automotive Engineering (IVK) and the Research Institute of Automotive Engineering and Vehicle Engines Stuttgart (FKFS). In particular, I would like to thank Dr. Michael Grill for promoting my interest on the topic, guiding my work and for his endless support. Many thanks to all the colleagues that contributed through countless fruitful discussions, especially T. Günther, Dr. M.-T. Keskin, K. Yang, L. Urban, C. Auerbach and P. Skarke. I will also not forget those unnamed here for providing me with an inspiring working environment.

I would also like to thank the working group and all the companies that supported the research tasks within the project "Knock with EGR at full load" defined and financed by the Research Association for Combustion Engines (FVV) e.V. My very sincere thanks also goes to my project partners M. Mally from the Institute for Combustion Engines of the RWTH Aachen University and Dr. L. Cai from the Institute for Combustion Technology of the RWTH Aachen University.

I would further like to express my deep sense of gratitude to Dr. A. Kulzer from Dr. Ing. h.c. F. Porsche AG for initiating and guiding the research project and for his continuous advice.

Finally yet importantly, I am very grateful to my family and my friends for their belief and support. They never doubted any of my decisions, they helped me to clear my mind in stressful times, and I am very glad to know that they will always stand by me. My special thanks go to my beloved one for her patience and the continuous support over the years.

Stuttgart   Alexander Fandakov

Contents

List of Figures

List of Tables

Nomenclature

Greek Letters

α	[°CA]	Crank Angle
α	[m^2/s]	Thermal Diffusivity of the Unburnt Gas in the Thermal Boundary Layer
α	[-]	Smoothing Factor
γ	[-]	Ratio of Specific Heats of Unburnt Gas in the Thermal Boundary Layer
δ_t	[m]	Boundary Layer Thickness at a Specified Cylinder Wall Location
ε	[-]	Wall Emissivity
η_c	[-]	Combustion Efficiency
κ_{spark}	[-]	Adiabatic Exponent at Spark
λ	[-]	Air-Fuel Equivalence Ratio
μ	[Ns/m^2]	Viscosity of Unburnt Gas in the Thermal Boundary Layer
μ_{EGR}	[Ns/m^2]	Viscosity of Exhaust Gas Fraction in the Thermal Boundary Layer
μ_{air}	[Ns/m^2]	Viscosity of Air Fraction in the Thermal Boundary Layer
ξ	[-]	Exponent For the Influence of Exhaust Gas
ρ	[kg/m^3]	Density of Unburnt Gas in the Thermal Boundary Layer
ρ_{ub}	[kg/m^3]	Unburnt Mixture Density
σ	[Wm^{-2}K^{-4}]	Stefan-Boltzmann Radiation Constant
τ	[s]	Ignition Delay of the Mixture at the Current Boundary Conditions
$\tau_{1,high}$	[s]	High-Temperature Ignition Delay in Low-Temperature Regime of Ignition

$\tau_{1,low}$	[s]	Low-Temperature Ignition Delay in Low-Temperature Regime of Ignition
$\tau_{2,high}$	[s]	High-Temperature Ignition Delay in Medium-Temperature Regime of Ignition
$\tau_{2,low}$	[s]	Low-Temperature Ignition Delay in Medium-Temperature Regime of Ignition
$\tau_{3,high}$	[s]	High-Temperature Ignition Delay in High-Temperature Regime of Ignition
τ_{high}	[s]	High-Temperature (Auto-) Ignition Delay
τ_L	[s]	Characteristic Burn-Up Time
τ_{low}	[s]	Low-Temperature Ignition Delay
υ	[m/s]	Gas Velocity at the Specified Cylinder Wall Location
υ_{Turb}	[m²/s]	Kinematic Turbulent Viscosity
φ	[°CA]	Crank Angle
φ_{ZS}	[-]	Parameter of the Cycle-to-Cycle Variations Model
χ_T	[-]	Pre-Factor
χ_{ZS}	[-]	Parameter of the Cycle-to-Cycle Variations Model
ϕ	[-]	Fuel-Air Equivalence Ratio
ϕ_1		Empirical Function for the Calculation of the Pertinent Reaction Product Concentration Change
ϕ_2		Empirical Function for the Calculation of the Pertinent Reaction Product Concentration Change

Roman Letters

A_{Fl}	[m²]	Flame Surface
$A_{i,high}$	[-]	Pre-Exponential Factor, High-Temperature Ignition Delay
$A_{i,low}$	[-]	Pre-Exponential Factor, Low-Temperature Ignition Delay
A_w	[m²]	Wall Surface Area

$B_{i,high}$	[K]	Activation Energy Parameter, High-Temperature Ignition Delay
$B_{i,low}$	[K]	Activation Energy Parameter, Low-Temperature Ignition Delay
C_1	[-]	Empirical Constant
C_1	[1/K^3]	Temperature Increase Model Parameter
C_2	[K]	Empirical Constant
C_2	[1/K^2]	Temperature Increase Model Parameter
C_3	[1/K]	Temperature Increase Model Parameter
C_4	[-]	Temperature Increase Model Parameter
C_5	[K]	Temperature Increase Model Parameter
C	[1/Pa^{C_1}]	Empirical Constant
C_k	[-]	Scaling parameter for the starting value of the specific turbulence
C_u	[-]	Isotropic turbulence speed scaling parameter
c_p	[J/kg/K]	Heat Capacity at Constant Pressure of Unburnt Gas in the Boundary Layer
c_v	[J/kg/K]	Heat Capacity at Constant Volume
D	[m]	Engine Bore
$\dfrac{dm_E}{dt}$	[kg/s]	Mass Flow into the Flame Zone (Mass Entrainment)
$\dfrac{dm_E}{d\varphi}$	[kg/°CA]	Exhaust Mass Flow
$\dfrac{dm_F}{d\varphi}$	[kg/°CA]	Flow Of Injected Fuel Mass
$\dfrac{dm_I}{d\varphi}$	[kg/°CA]	Inlet Mass Flow
$\dfrac{dm_L}{d\varphi}$	[kg/°CA]	Leakage Mass Flow (Blowby)
$\dfrac{dm_b}{dt}$	[kg/s]	Mass Flow into the Burnt Zone
$\dfrac{dm_{ub}}{dt}$	[kg/s]	Mass Flow into the Unburnt Burnt Zone

$\dfrac{dm}{dt}$	[kg/s]	Total Mass Change
$\dfrac{dm}{d\varphi}$	[kg/°CA]	Total Mass Flow / Cylinder Mass Change
$\dfrac{dp_{cyl}}{d\varphi}$	[bar/°CA]	Cylinder Pressure Change
$\dfrac{dQ_B}{d\varphi}$	[J/°CA]	Heat Release Rate
$\dfrac{dQ_W}{d\varphi}$	[J/°CA]	Wall Heat Flux
$\dfrac{dR}{d\varphi}$	[J/kg/K/°CA]	Individual Gas Constant Change
$\dfrac{dT}{dt}$	[K/s]	System Temperature Change
$\dfrac{dT}{d\varphi}$	[K/°CA]	Temperature Change
$\dfrac{dU}{dt}$	[J/s]	Total Internal Energy Change
$\dfrac{dU}{d\varphi}$	[J/°CA]	Internal Energy Change
$\dfrac{dV}{dt}$	[m³/s]	Volume Change
$\dfrac{dV}{d\varphi}$	[m³/°CA]	Volume Change
$\dfrac{dY_k}{dt}$	[1/s]	Change of Species Mass Fraction
$\dfrac{d\varphi}{dt}$	[°CA/s]	Change of Crank Angle over Time
F_1		Empirical Function for the Calculation of the Pre-Exponential Factor, High-Temperature Ignition Delay
F_2		Empirical Function for the Calculation of the Activation Energy Parameter, High-Temperature Ignition Delay

F	[-]	Fuel-Air Ratio
f_1		Empirical Function for the Calculation of the Pre-Exponential Factor, Low-Temperature Ignition Delay
f_2		Empirical Function for the Calculation of the Activation Energy Parameter, Low-Temperature Ignition Delay
f_w	[-]	Constant for Wall Facing
g_1		Empirical Function for the Calculation of the Parameter C_1, Temperature Increase Resulting from the First Ignition Stage
g_2		Empirical Function for the Calculation of the Parameter C_2, Temperature Increase Resulting from the First Ignition Stage
g_3		Empirical Function for the Calculation of the Parameter C_3, Temperature Increase Resulting from the First Ignition Stage
g_4		Empirical Function for the Calculation of the Parameter C_4, Temperature Increase Resulting from the First Ignition Stage
g_5		Empirical Function for the Calculation of the Parameter C_5, Temperature Increase Resulting from the First Ignition Stage
H_u	[J/kg]	Lower Heating Value, LHV
h_E	[J/kg]	Specific Exhaust Enthalpy
h_I	[J/kg]	Specific Inlet Enthalpy
h_{in}	[J/kg]	Specific Enthalpy of Species Entering the System Through Inlets
h	[J/kg]	Specific Enthalpy
I_k	[-]	Pre-Reaction State of the Air-Fuel Mixture
i	[-]	Temperature Regime Index
i	[-]	Calculation Step Index
K	[m/s/Pa]	Non-Negative Constant

k	[-]	Species Index
k	[W/mK]	Thermal Conductivity of Unburnt Gas in the Thermal Boundary Layer
l_T	[m]	Taylor Length
l	[m]	Integral Length Scale
MON_{Eth}	[-]	Motor Octane Number of Ethanol
MON	[-]	Surrogate Motor Octane Number
M_X	[kg/kmol]	Mole Mass of Component X
$\dot{m}_0$	[kg/s]	Mass Flow Specified as a Constant or a Function of Time
$\dot{m}_X$	[kg/h]	Mass Flow of Component X
$\dot{m}_{in}$	[kg/s]	Mass Flow Through Inlets
$\dot{m}_{k,gen}$	[kg/s]	Mass Flow of Generated Species
$\dot{m}_{out}$	[kg/s]	Mass Flow Through Outlets
$\dot{m}_{wall}$	[kg/s]	Production of Homogeneous Phase Species on the Walls
$m_{cyl.total}$	[kg]	Total Mass of Components in Cylinder
$m_{exh.gas}$	[kg]	Cylinder Exhaust Gas Mass
m_F	[kg]	Flame Zone Mass
m_f	[kg]	Cylinder Fuel Mass
m_{fuel}	[kg]	Cylinder Fuel Mass
m_{water}	[kg]	Cylinder Water Mass
m	[kg]	Mass
n	[min^{-1}]	Engine Speed
Pr	[-]	Prandtl Number at the Specified Cylinder Wall Location
p_{cyl}	[Pa]	Cylinder Pressure
p_{incr}	[bar]	Pressure Increase Resulting from Low-Temperature Ignition
p_{ub}	[bar]	Unburnt Pressure
p	[Pa]	Pressure

$\dot{Q}$	[J/s]	Total Rate of Heat Transfer Through All Walls
q_0	[W/m^2]	Heat Flux Specified as a Function of Time
R	[J/kg/K]	Individual Gas Constant
Re	[-]	Reynolds Number at the Specified Cylinder Wall Location
RON_{Eth}	[-]	Research Octane Number of Ethanol
RON	[-]	Surrogate Research Octane Number
s_L	[m/s]	Laminar Flame Speed
s	[m]	Engine Stroke
$T_{KS,FTDC}$	[K]	Knock-Spot Temperature at Firing Top Dead Center
T_{bl}	[K]	Boundary Layer Temperature
$T_{incr,fit}$	[K]	Modeled Temperature Increase Resulting from Low-Temperature Ignition
T_{incr}	[K]	Temperature Increase Resulting from Low- Temperature Ignition
T_{low}	[K]	Temperature at time of Low-Temperature Ignition
T_{off}	[K]	Temperature Offset Representing the Knock-Spot
$T_{ub,spark}$	[K]	Unburnt Temperature at Spark
T_{ub}	[K]	Unburnt Mixture Temperature
T_{wall}	[K]	Cylinder Wall Temperature
T	[K]	Temperature
t_1	[s]	Predicted Time of Low-Temperature Ignition
t_2	[s]	Predicted Time of High-Temperature (Auto-) Ignition
t_e	[s]	Time at End of Integration / Overall Auto-Ignition Reaction Time
t	[s]	Time
t	[s]	Time Elapsed Since the Beginning of the Thermal Boundary Layer Development at the Specified Cylinder Wall Location
U	[J]	Total Internal Energy

U	[W/m²/K]	Heat Transfer Coefficient
u_E	[m/s]	Speed of the Flame Front Penetrating the Unburned Zone
u_{Turb}	[m/s]	Isotropic Turbulence Speed
u_k	[J/kg]	Specific Internal Energy of Species K
$V\dot{\omega}_k W_k$	[kg/s]	Generation Rate
$V_{cyl,FTDC}$	[m³]	Cylinder Volume at Firing Top Dead Center
$V_{cyl,spark}$	[m³]	Cylinder Volume at Spark
$V_{ub,bl}$	[-]	Unburnt Volume Fraction in the Thermal Boundary Layer
V	[m³]	Volume
v_0	[m/s]	Velocity Specified as a Function of Time
v_C	[-]	Volume Fraction of Component C
v_p	[m/s]	Piston Velocity
v	[m/s]	Wall Velocity
w	[-]	Wall Index
X_{Eth}	[-]	Mole Fraction of Ethanol
x_{Eth}	[-]	Ethanol Mass Fraction
$x_{EGR,st}$	[-]	Stoichiometric Exhaust Gas Recirculation Rate
x_{EGR}	[-]	EGR Mass Fraction
x_{Hep}	[-]	N-Heptane Mass Fraction
x_{Iso}	[-]	Iso-Octane Mass Fraction
x_{Tol}	[-]	Toluene Mass Fraction
x_o	[m]	Coordinate of the Specified Cylinder Wall (Piston / Liner) Location
$x_{ub,bl,norm}$	[-]	Normalized Unburnt Mass Fraction in Thermal Boundary Layer
$x_{ub,bl,smth}$	[-]	Smoothed Unburnt Mass Fraction in Thermal Boundary Layer
$x_{ub,bl}$	[-]	Unburnt Mass Fraction in the Thermal Boundary Layer

x	[mol/m³]	Concentration of Pertinent Reaction Products
x	[m]	Current Distance between the Piston Top and the Cylinder Head
Y_X	[-]	Volume Fraction of Component X
$Y_{k,in}$	[-]	Mass Fractions of Species Entering the System Through Inlets
Y_k	[-]	Species Mass Fraction

Subscripts

Air, air	Air
B	Burn, Combustion
b	Burnt
bl	Boundary Layer
c	Combustion
CO₂	Carbon Dioxide
cyl	Cylinder
E	Exhaust
E	Entrainment
e	End
EGR	Exhaust Gas Recirculation
Eth	Ethanol
etha	Ethanol
Exh, exh	Exhaust
F, f	Fuel
fit	Fitted, Modelled
Fl	Flame
FTDC	Firing Top Dead Center
gen	Generated
grad	Gradient
Hep	n-Heptane

high	High-Temperature (Auto-) Ignition
I	Inlet / Intake
incr	Increase
Int	Intake
iO	iso-Octane
Iso	iso-Octane
KS	Knock-Spot
L	Leakage
L	Laminar
lam	Lambda
low	Low-Temperature Ignition
norm	Normalized
off	Offset
out	Outlet
p	Piston
smooth	Smoothed
spark	Crank Angle of Spark Timing
st	Stoichiometric
T	Taylor
t	Thermal
Tol, tol	Toluene
Turb	Turbulence
ub	Unburnt
W, w	Wall
wall	Wall
wat	Water
ZS	„*Zyklenschwankung*"[1], Cycle-to-Cycle Variation

---

[1] Term in German.

Acronyms

AFR	Air-Fuel Equivalence Ratio
aFTDC	After Firing Top Dead Center
aTDC	After Top Dead Center
AWC	Averaged Working Cycle
BC	Boundary Conditions
bTDC	Before Top Dead Center
CA	Crank Angle
CCV	Cycle-to-Cycle Variations
CFD	Computational Fluid Dynamics
CO	Carbon Monoxide
CO_2	Carbon Dioxide
COV	Coefficient of Variance
D	Dimensional
DI	Direct Injection
DOHC	Double Overhead Camshaft
EB	Energy Balance
EGR	Exhaust Gas Recirculation
FTDC	Firing Top Dead Center
gHCCI	gasoline Homogeneous Charge Compression Ignition
H_2	Hydrogen
HC	Hydrocarbons
IMEP	Indicated Mean Effective Pressure
IMPG	Integral of the Modulus of the Pressure Gradient
IMPO	Integral of the Modulus of Pressure Oscillations
IVC	Inlet Valve Close
KLSA	Knock Limited Spark Advance
KO	Knock Onset
KPP	Knock Peak-to-Peak
LES	Large Eddy Simulation

LHV	Lower Heating Value
MAPO	Maximum Amplitude of the Pressure Oscillations
MFB	Mass Fraction Burnt
MON	Motor Octane Number
NO	Nitric Oxide
NO_x	Nitrogen Oxides[2]
NTC	Negative Temperature Coefficient
O_2	Oxygen
ODE	Ordinary Differential Equations
OH	Hydroxyl Radical
PAH	Polycyclic Aromatic Hydrocarbon
PN	Particulate Number
PTA	Pressure Trace Analysis
RANS	Reynolds-Averaged Navier–Stokes
RCM	Rapid Compression Machine
RON	Research Octane Number
RPM	Revolutions per Minute
SEPO	Signal Energy of the Pressure Oscillations
SI	Spark Ignition
ST	Shock Tube
SWC	Single Working Cycle
TDC	Top Dead Center
TPA	Three Pressure Analysis
VCR	Variable Compression Ratio

---

[2] NO_x is a generic term for nitric oxide (NO) and nitrogen dioxide (NO_2) commonly used in the context of internal combustion engine emissions.

Abstract

The enforcement of lower fuel consumption as well as the tightening emission standards, together with the requirements posed to spark ignition engines by the powertrain hybridization, imposes a significant improvement of the efficiency over the entire engine map. New concepts are needed to guarantee the clean and efficient engine operation in a very wide range of operating conditions, especially at high loads, where knock typically occurs.

Nowadays, the 0D/1D simulation of internal combustion engines is commonly used in the engine concept design phase. Thanks to the high prediction quality of the phenomenological models and the low computational times, this is a powerful tool used to reduce development costs by partially eliminating the need for cost-intensive test bench investigations. However, the existing 0D/1D models for predicting knock, which are commonly based on the Livengood-Wu integral, are known for their poor performance and the great effort needed for their calibration. This fact results in significant restrictions on the development of future spark ignition engine concepts within a 0D/1D simulation environment, as a reliable, fully predictive knock model is an essential requirement for accomplishing this task.

In this work, reaction kinetic simulations of measured knocking single cycles at in-cylinder conditions are performed by using a detailed reaction kinetics mechanism with a model representing the unburnt zone of a two-zone spark ignition combustion model. The investigations show that, at specific operating conditions, the auto-ignition in the unburnt mixture that precedes the occurrence of knock happens in two stages. In this case, low-temperature ignition occurs in the unburnt mixture while the combustion is taking place. This phenomenon significantly influences the auto-ignition behavior of the mixture, thus severely impairing the prediction capabilities of the knock integral that all commonly used 0D/1D knock models are based on. Hence, an improved approach for modeling the progress of the chemical reactions is needed for the accurate prediction of the knock boundary.

Based on these findings, a new phenomenological two-stage approach reproducing the auto-ignition behavior of the detailed mechanism at in-cylinder

conditions is developed in this work. The occurrence of each of the two ignition events is predicted by a single integral. The inputs of the two coupled
integrals are the values of the ignition delay for the corresponding ignition
stage as a function of the current boundary conditions. For this purpose, an
enhanced three-zone approach for modeling the influence of various parameters (pressure, temperature, exhaust gas, air-fuel equivalence ratio, ethanol and
water content as well as surrogate composition) on the auto-ignition delay
times of the mixture is developed. Furthermore, models for the delay of the
low-temperature ignition as well as the temperature increase resulting from the
first ignition stage as a function of the boundary conditions are formulated.
Finally, it is demonstrated that the novel two-stage auto-ignition model predicts the occurrence of two-stage ignition and considers the significant influence of low-temperature heat release on the mixture's auto-ignition behavior
very accurately at various operating conditions.

However, the correct prediction of local auto-ignition is not sufficient for the
reliable simulation of the knock boundary, as the occurrence of this phenomenon does not necessarily result in knock. Except for not considering low-temperature ignition, commonly used knock models assume that no knock can
occur after a pre-defined, constant mass fraction burnt point. The evaluation
of the measured knocking single cycles however shows that the latest possible
mass fraction burnt point where knock can occur fluctuates because of the cycle-to-cycle variations and changes significantly with parameters such as engine speed, exhaust gas recirculation rate and the air-fuel equivalence ratio.
Hence, a cycle-individual criterion for occurrence of knock considering the
current operating conditions is needed. To this end, an approach based on the
unburnt mass fraction in the thermal boundary layer at the time of auto-ignition
is proposed. The boundary layer volume is estimated with a phenomenological
model and, because of the cool cylinder walls, it has a temperature that is much
lower than the mean unburned mass temperature. Besides the operating conditions, the developed knock occurrence criterion also accounts for the flame
propagation and the cylinder geometry. It is assumed that if the unburnt mass
fraction in the boundary layer at the predicted time of auto-ignition is higher
than a pre-defined threshold calibrated at the measured knock boundary, no
knock can occur.

The new knock model contains no empirical measurement data fits and has
just one engine-specific calibration parameter that does not depend on the op-

erating conditions. Therefore, the new model can be applied to different engines without any limitations. Finally, an extensive model validation against measurement data on different engines at various operating conditions is performed. A comparison with today's industry standards reveals the huge gain in knock boundary prediction accuracy achieved in the present work. Thus, the new knock model contributes substantially to an efficient development process of future spark ignition engine concepts within a 0D/1D simulation environment.

Zusammenfassung

Das Bestreben nach einem minimierten Kraftstoffverbrauch sowie die Verschärfung der Abgasnormen, zusammen mit den Anforderungen an fremdgezündeten Verbrennungsmotoren durch die Hybridisierung des Antriebsstrangs, erfordern die signifikante Verbesserung des Wirkungsgrades im gesamten Motorkennfeld. Es sind neue Ottomotorkonzepte erforderlich, um den sauberen und effizienten Motorbetrieb zu gewährleisten, insbesondere bei hohen Lasten, bei denen typischerweise Klopfen auftritt.

Die 0D/1D-Simulation von Verbrennungsmotoren ist ein mächtiges Werkzeug, das heutzutage verstärkt in der Motorkonzeptentwicklungsphase eingesetzt wird. Dank der hohen Vorhersagegüte der phänomenologischen Modelle und der geringen Rechenzeiten stellt die Arbeitsprozessrechnung ein leistungsfähiges Tool zur Reduzierung der Entwicklungskosten dar, indem teure Motorprüfstandsuntersuchungen teilweise entfallen. Die vorhandenen 0D/1D-Modelle zur Vorhersage der Klopfgrenze, die größtenteils auf dem Livengood-Wu Integral basieren, sind allerdings für ihre eingeschränkte Vorhersagefähigkeit sowie großen Kalibrierungsaufwand bekannt. Diese Tatsache führt zu erheblichen Einschränkungen bei der Entwicklung zukünftiger Ottomotorkonzepte in einer 0D/1D-Simulationsumgebung, da ein zuverlässiges, vollprädiktives Klopfmodell eine wesentliche Voraussetzung für die Erfüllung dieser Aufgabe ist.

Die in der Arbeitsprozessrechnung verwendeten Klopfmodelle berechnen den Vorreaktionszustand des unverbrannten Gemisches mithilfe des Klopfintegrals, das einen vereinfachten Ansatz für die Auswertung des Reaktionsfortschritts im Endgas darstellt. Die in dieser Arbeit durchgeführten reaktionskinetischen Simulationen der unverbrannten Zone eines zweizonigen Verbrennungsmodells zeigen, dass bei bestimmten Randbedingungen die dem Klopfen vorausgehende Selbstzündung im Unverbrannten in zwei Stufen ablaufen kann. Hierbei findet eine von der Verbrennung überdeckte Niedertemperaturwärmefreisetzung statt, die einen signifikanten Einfluss auf die Selbstzündung im unverbrannten Gemisch hat und somit die Vorhersagefähigkeit des Klopfintegrals stark einschränkt. Für die genaue Vorhersage der Selbstzündung ist deshalb die Entwicklung eines neuen Modells, das die Zweistufenzündung genau beschreibt, erforderlich.

Ausgehend von diesen Erkenntnissen wird in dieser Arbeit ein neuer, zweistufiger Ansatz für die Vorhersage der Selbstzündung im Unverbrannten entwickelt. Das Auftreten beider Zündstufen wird hierbei von jeweils einem Integral vorhergesagt. Die Eingänge der zwei verknüpften Integrale sind die Zündverzugszeiten der jeweiligen Zündstufe als Funktion der momentanen Randbedingungen. Zu diesem Zweck wird ein dreizoniger Modellierungsansatz erarbeitet, der die Einflüsse von Temperatur, Druck, Verbrennungsluftverhältnis, rückgeführtem Abgas, eingespritztem Wasser sowie Surrogatzusammensetzung auf den Verzug der Selbstzündung abbildet. Des Weiteren werden geeignete Modelle für den Zündverzug der Niedertemperaturzündung sowie den Temperaturanstieg, der aus der Niedertemperaturwärmefreisetzung resultiert, als Funktion der Randbedingungen formuliert. Das zweistufige Selbstzündungsmodell berechnet das Auftreten der Zweistufenzündung und beschreibt den signifikanten Einfluss der ersten Zündstufe auf das Selbstzündungsverhalten des Gemisches bei unterschiedlichen Motorbetriebsbedingungen.

Allerdings resultiert eine Selbstzündung im Unverbrannten nicht unbedingt in Klopfen. Somit ist für die verlässliche Vorhersage der Klopfgrenze zusätzlich ein Klopfkriterium erforderlich. Die in der Arbeitsprozessrechnung verwendeten Klopfmodelle basieren auf der Annahme, dass nach einem bestimmten, konstanten Umsatzpunkt kein Klopfen auftreten kann. Die Auswertung gemessener Einzelarbeitsspiele zeigt allerdings, dass dieser Umsatzpunkt aufgrund der Zyklenschwankungen sowie bei Änderungen der Motorbetriebsbedingungen stark variiert, was ein arbeitsspielindividuelles Klopfkriterium notwendig macht. Zu diesem Zweck wird in dieser Arbeit ein Ansatz erarbeitet, der auf dem unverbrannten Massenanteil in der thermischen Grenzschicht basiert. Hierzu wird ein phänomenologisches Modell für das Grenzschichtvolumen erstellt, welches die Motorbetriebsbedingungen, Brennraumgeometrie sowie die Flammenausbreitung berücksichtigt. Die Temperatur in der Grenzschicht ist aufgrund der kalten Zylinderwände erheblich niedriger als die Temperatur im Unverbrannten. Daher wird für die Klopfvorhersage angenommen, dass im Fall eines unverbrannten Massenanteils in der Grenzschicht zum Zeitpunkt der vorhergesagten Selbstzündung, der über einem vordefinierten Schwellwert liegt, trotz der Selbstzündung kein Klopfen auftreten kann. Dieser Schwellwert ist der einzige Abstimmparameter des neuen Klopfmodells. Er ist motorspezifisch, unabhängig von den Motorbetriebsbedingungen und muss anhand eines Betriebspunktes an der gemessenen Klopfgrenze kalibriert werden.

Der entwickelte Klopfmodellierungsansatz beinhaltet keine empirischen Messdatenfits und kann somit ohne Einschränkungen auf unterschiedliche Motorkonfigurationen angewendet werden. Das Modell wird abschließend durch Abgleich von Simulationsergebnissen mit Messdaten von mehreren Motoren validiert. Der durgeführte Vergleich mit den heutigen Industriestandards offenbart die erzielte signifikante Steigerung der Güte der Klopfvorhersage. Das vollprädiktive Klopfmodell leistet somit einen entscheidenden Beitrag zu einer kosteneffizienten Entwicklung zukünftiger Ottomotorkonzepte in der Arbeitsprozessrechnung.

1 Introduction

The ongoing trend of enforcing lower fleet consumption and tightening emission standards, combined with ever-increasing customer demands and severe competition on the global market is the main driving force for further improvement of the internal combustion engine regarding its efficiency and emissions. Additional challenges are posed by the ongoing powertrain hybridization, as it imposes a significant improvement of the efficiency over the entire engine map [103]. Hence, new concepts are needed to guarantee the clean and energy-saving engine operation in a very wide range of operating conditions, especially at high loads, where knock typically occurs [122]. This phenomenon still represents the most significant operation limit of spark ignition (SI) engines, although knock has been observed for the first time almost 100 years ago [75].

The main theory behind engine knock is that a detonation is being caused by auto-ignition(s), thus resulting in extremely rapid heat release in the end gas ahead of the propagating turbulent flame [89]. This causes very high cylinder pressures and temperatures that could potentially lead to total engine failure [75]. High compression ratios, which are desired for further improving efficiency, result in increased cylinder pressures and temperatures that, in turn, promote the occurrence of knock. Thus, the auto-ignition phenomenon is a major obstacle in increasing the indicated engine efficiency. This is particularly important for turbocharged and downsized direct injection SI engines because of the higher inlet and, hence, cylinder pressures [153].

Nowadays, the 0D/1D simulation of internal combustion engines is commonly used in the engine concept design phase. Thanks to the high prediction quality of the phenomenological models and the low computational times, this is a powerful tool used to reduce development costs by partially eliminating the need for cost-intensive test bench investigations [52] [134]. However, the existing 0D/1D models for predicting knock, which are commonly based on the Livengood-Wu integral [105], are known for their poor performance and the great effort needed for their calibration [52] [53]. Hence, the reliable prediction of the knock boundary is a challenging task, especially if the entire engine map has to be considered. This fact results in significant restrictions on the

© Springer Fachmedien Wiesbaden GmbH, part of Springer Nature 2019
A. Fandakov, *A Phenomenological Knock Model for the Development of Future Engine Concepts*, Wissenschaftliche Reihe Fahrzeugtechnik Universität Stuttgart, https://doi.org/10.1007/978-3-658-24875-8_1

development of future SI engine concepts within a 0D/1D simulation environment, as a reliable, fully predictive knock model is an essential requirement for accomplishing this task.

Besides the state of the art that involves direct fuel injection, variable valve timing and lift, water-cooled intercoolers as well as combustion process and engine cooling optimization (e.g. integrated exhaust manifolds), technologies for reducing the knock propensity such as high load exhaust gas recirculation (EGR), lean combustion and water injection emerge as new, challenging tasks in the 0D/1D SI engine simulation. In the following section, some of the key concepts for suppressing knock that could be employed within the framework of future SI engine concepts are highlighted.

1.1   Key Technologies for Knock Suppression

The variable compression ratio (VCR) is a concept nearly as old as the internal combustion engine itself [83]. It represents a very beneficial and flexible technology for reducing the fuel consumption of internal combustion engines because of the direct relationship between the compression ratio and the engine efficiency [138]. Numerous mechanical, hydraulic and electric VCR concepts have been developed and patented over the past decades. The most feasible designs are based on a continuously or stepwise variable connecting rod length. However, the inevitable increase of the friction losses has to be taken into account when comparing VCR with conventional engine operation. The major disadvantage of this technology though is its complexity that also results in high costs.

Alternatively, the Miller cycle can be applied. It is characterized by a gas exchange with shortened intake valve timing with the primary goal of an over-expansion during the expansion stroke. This state-of-the-art technology [83] leads to an improved conversion of the hot gas expansion work into mechanical work at the crankshaft [122]. Additionally, the reduced length of the inlet valve opening event at full load results in a lower effective compression ratio. Thus, both the peak temperature and pressure at the end of the compression stroke decrease, which has a positive effect on the knock propensity of the engine. Hence, Miller events enable a more efficient engine operation (advanced spark timing) for the same geometric compression ratio [111]. However, it should be remarked that this concept has been reported to negatively influence the in-cylinder charge motion [128]. Therefore, the implementation of Miller events generally requires the application of additional technologies like valve masking in order to maintain the level of the in-cylinder charge motion [170].

The Miller concept can be combined with the recirculation of exhaust gas. EGR is a state of the art technology that has been used in diesel engines for several years with focus on the reduction of the NO_x emissions [83]. With a different aim, but through similar mechanisms, cooled external EGR becomes relevant for modern gasoline applications as well. As the conventional SI combustion process is controlled by adjusting the air-fuel mixture quantity (at a constant air-fuel equivalence ratio) [122], the part load operation is characterized by high flow losses at the throttle body that controls the mixture amount entering the cylinders and thus the engine load. Because the charge dilution with exhaust gas reduces the air and fuel mass fractions in the mixture, EGR enables the increase of the cylinder charge without changing the engine load, thus reducing the flow losses at the throttle body, which is known as de-throttling.

At high and full load, the dilution effect of EGR becomes a disadvantage, as higher charge pressures and consequently a larger turbocharger is required for maintaining the same engine load level. On the other hand, diluting the air-fuel mixture with exhaust gas in this case increases the heat capacity of the cylinder charge, which results in lower peak temperature and pressure at the end of the compression stroke [111]. Furthermore, the exhaust gas leads to slower combustion [71], so that the temperature of the unburnt mixture during the combustion declines as well. Finally yet importantly, the mixture dilution with exhaust gas slows down the chemical reactions leading to auto-ignition in the unburnt mixture [26]. Taken together, these effects of EGR result in a reduced knock propensity and an engine efficiency increase. However, recent studies have revealed that the knock mitigation effect of EGR strongly depends on the engine load [145].

Besides the Miller cycle, cooled exhaust gas recirculation, and variable compression ratio, the injection of water has recently gained increased attention as a promising technology for significant CO_2 reduction. As water dilutes the air-fuel mixture, the effects of this technology on the occurrence of knock are similar to those of EGR: higher heat capacity of the cylinder charge (thermo-dynamic effect) as well as a slower auto-ignition process in the unburnt mixture (chemical effect). The potential of water injection to reduce the fuel consumption by suppressing knock and thus allowing for earlier spark timing at the knock limited spark advance increases with the engine load demand. However, the stated positive effects of this concept are closely associated with a rather high water consumption [78]. This requires the careful optimization of the trade-off between fuel and water demand. In this context, water supply technologies like the condensation of water via air conditioning [83] or from the exhaust gas [12] are of interest. A further disadvantage of water injection is the required hardware and its costs. Water injection can also be used in combination with the Miller valve timing and high load EGR [78] [83].

A further promising technology for achieving high SI engine efficiency is the super-lean burn concept, which again involves mixture dilution, in this case by excess air (air-fuel equivalence ratios of up to 2). This increases the heat capacity of the mixture and leads to very long combustion durations [71] that result in low unburnt mixture temperatures and wall heat losses reduction. Furthermore, the excess air slows down the mixture auto-ignition process [26]. Thus, this technology is a promising approach for mitigating knock. However, the super-lean combustion is also very unstable, as the cycle-to-cycle variations increase dramatically in this case [161]. This requires a combustion stabilization by enhancing the in-cylinder flow, e.g. by adopting a long stroke for the generation of high tumble flow and an optimization of the intake port shape, as well as a high-energy ignition system [171]. Furthermore, the excess air and thus oxygen in the exhaust gas imply that the commonly used three-way catalytic converter cannot be applied in combination with this technology [122]. Nevertheless, resent studies have revealed that indicated thermal efficiencies of over 45 % can be achieved with the super-lean burn concept [171].

1.2   Motivation and Objectives

The 0D/1D simulation is an essential approach for the cost-effective development of clean and efficient spark ignition engines. However, the available commonly used knock models are characterized by poor prediction performance and high calibration effort [53]. The review of the key technologies for future high-efficiency SI engine concepts in the previous section reveals that a 0D/1D knock model has to be capable of predicting the influence of numerous parameters on the knock behavior, such as engine load and compression ratio, EGR, valve timing, inlet temperature, excess air or fuel as well as injected water. A predictive 0D/1D knock model further has to consider the well-known fuel, engine speed and combustion chamber geometry effects on the knock propensity [168].

Therefore, the main objective of this work is the development of a new phenomenological, fully predictive 0D/1D knock model that accounts for the effects of all parameters influencing knock occurrence, thus enabling the investigation of future SI engine concepts in a 0D/1D simulation environment.

To begin with, Chapter 2 presents the fundamentals of the 0D/1D engine simulation and reviews the existing knock modeling approaches. The effects of factors known to influence knock from previous studies are analyzed in Chapter 3. Here, the commonly used knock modeling approach is applied to measured single cycles, aiming at identifying reasons for its well-known poor prediction performance. Additionally, the test bench setup used for the experimental investigations of knock occurrence is presented and the processing of the measurement data is discussed.

As it is commonly assumed that knock occurs because of local auto-ignition(s) in the unburnt mixture, the accurate prediction of this phenomenon is the key to developing a fully predictive knock model for the 0D/1D engine simulation. For this reason, Chapter 4 aims at better understanding the chemical processes resulting in auto-ignition and developing an appropriate approach for predicting the occurrence of this phenomenon. To this end, investigations with a detailed reaction kinetics mechanism at in-cylinder conditions are performed and the weaknesses of the commonly used knock integral are identified and examined in detail. Subsequently, a new auto-ignition prediction approach considering the effects of all currently conceivable knock suppression measures that could be employed within the framework of future SI engine concepts is developed and validated.

However, the correct prediction of local auto-ignition is not sufficient for the reliable simulation of the knock boundary, as the occurrence of this phenomenon does not necessarily result in knock. Therefore, a phenomenological knock criterion based on the unburnt mass fraction in the thermal boundary layer at the predicted time of auto-ignition is proposed in Chapter 5, to account for the well-known knock probability decrease towards the end of combustion.

Chapter 6 reviews the possible sources for the model inputs needed for the simulation of the knock boundary and evaluates the prediction quality they result in. Additionally, investigations are performed to ensure that the developed approach can be used in combination with various other simulation models. Subsequently, the knock model workflow, outputs, as well as the calibration process are described in detail. Finally, in Chapter 7 the newly developed 0D/1D approach is applied to available measurement data and its prediction performance is evaluated extensively at various operating conditions on different engines. The conclusive comparison of the knock boundary simulation quality achieved with the new model and two state-of-the-art commercial knock models demonstrates the huge accuracy gain achieved in this work.

2 Fundamentals and State of the Art

Nowadays 0D/1D simulations are being widely used in the engine development process. Thanks to the high prediction quality of the models and the low
computational times, this is a powerful tool used to reduce development costs
by partially eliminating the need for cost-intensive test bench investigations.
This chapter presents the fundamentals of the 0D/1D engine simulation and
reviews the existing knock modeling approaches. However, first the basics of
spark ignition combustion, abnormal combustion phenomena, and kinetic
modeling shall be briefly discussed.

2.1 Spark Ignition Combustion

In conventional spark ignition engines, fuel, air and residual gas are mixed
together and then compressed. Under normal operating conditions, combustion is initiated towards the end of the compression stroke at the spark plug by
an electric discharge[3]. Right after the spark discharge, the energy release from
the developing flame is too small for a pressure and temperature rise due to
combustion to be discerned. Following the inflammation, a turbulent flame
develops and propagates through the fuel-air-burned gas mixture. As the flame
grows and propagates across the combustion chamber, the pressure steadily
rises above the value it would have in the absence of combustion. Usually, the
cylinder pressure and temperature reach their maxima after top dead center,
but before the cylinder charge is fully burned [122]. Then they decrease, as the
cylinder volume continues to increase during the expansion stroke. As the
combustion progresses, the propagating flame reaches the combustion chamber walls and then extinguishes.

---

[3] Details on the stratified and homogeneous charge compression ignition (HCCI) combustion processes, which are not discussed in this work, can be found in [75] [85] [153].

© Springer Fachmedien Wiesbaden GmbH, part of Springer Nature 2019
A. Fandakov, *A Phenomenological Knock Model for the Development of
Future Engine Concepts*, Wissenschaftliche Reihe Fahrzeugtechnik Universität
Stuttgart, https://doi.org/10.1007/978-3-658-24875-8_2

The flame development and subsequent propagation vary from cycle to cycle, because the flame growth depends on the local mixture motion and composition [75]. These quantities vary in successive cycles in any given cylinder and may also change on a cylinder-to-cylinder basis. Especially important are the mixture motion and composition near the spark plug at the time of spark discharge, as these govern the early stages of flame development [161]. Hence, the cycle-to-cycle combustion fluctuation is a phenomenon typical for spark ignition engines that is crucial because the "extreme" cycles limit the operating range of the engine [75].

However, a handful of factors, e.g. the fuel composition, engine design, operating conditions and combustion chamber deposits, may prevent the combustion process from taking place as described above ("normal" combustion) [156]. In this case, abnormal combustion occurs. Two major types of abnormal combustion exist: knock and surface ignition [75]; however, knock is by far the most important abnormal combustion phenomenon for reasons that shall be discussed in detail in the following sections.

2.1.1   Abnormal Combustion

As the flame propagates across the combustion chamber, the unburned mixture ahead of the flame (end gas) is compressed, causing its pressure, temperature, and density to increase. Thus, the chemical reactions taking place in the unburnt mixture are accelerated [166]. Consequently, an auto-ignition characterized by a rapid combustion reaction that is not initiated by any external ignition source might occur and lead to a rapid release of chemical energy [168]. When this happens, the end gas burns very rapidly, releasing its energy at a rate 5 to 25 times that characteristic of normal combustion [75]. This causes high-frequency pressure oscillations inside the cylinder resulting from the propagation of pressure waves of substantial amplitude across the combustion chamber, Figure 2.1, that produce the sharp metallic noise known as "knock". The occurrence of knock reflects the outcome of a race between the propagating flame front and the speed of the chemical reactions in the end gas that are influenced by the boundary conditions in the unburned mixture, e.g. temperature and pressure [75]. No abnormal combustion will occur if the flame front consumes the end gas before the chemical reactions have caused the air-fuel mixture to auto-ignite.

Surface ignition on the other hand results in uncontrolled combustion and is caused by any means other than the normal spark discharge ("hot-spot"), e.g. an overheated valve or spark plug or a glowing combustion chamber deposit on the combustion chamber walls [166]. It can occur before spark (pre-ignition) or after it (post-ignition) [75]. Following the surface ignition, a turbulent flame develops at each ignition location and starts to propagate across the chamber in an analogous manner to what occurs with normal spark ignition [49]. Uncontrolled combustion is most evident and its effects most severe when it results from pre-ignition [75]. However, even if a surface ignition occurs after the spark plug fires (post-ignition), the spark discharge no longer has the complete control over the combustion process. It should be remarked that surface ignition might result in knock [166]. However, the occurrence of surface ignition is a problem that can usually be solved by paying special attention to the engine design as well as the fuel and especially the lubricant quality [41].

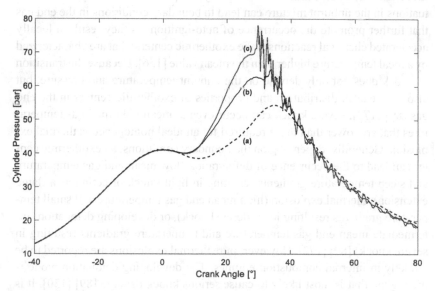

Figure 2.1: Unfiltered cylinder pressure trace of single engine cycles. (a) normal SI combustion, (b) light knock and (c) severe knock.

Auto-ignition of the unburned mixture in the end gas zone leading to engine knock occurs when the temperature and pressure in the unburned zone in front of the flame are sufficiently high and enough time is available for the chemical reactions to take place (known as induction time). Therefore, engine designs and operating conditions leading to critical boundary conditions in the end gas in respect of auto-ignition (high unburnt temperatures and pressures) promote the occurrence of knock. For this reason, the phenomenon is often related to high compression ratios and charge densities as well as early spark timings. Hence, knock is an inherent constraint on engine performance and efficiency since it limits decisive parameters like the maximum compression ratio that can be used with any given fuel as well as the earliest possible combustion center position [75].

During the combustion, the hot burnt gases behind the regular flame that is triggered by the spark plug lead to an in-cylinder pressure increase. Clearly, the still unburnt air-fuel mixture that is in front of the propagating flame is exposed to the higher pressure and thus temperature too, which reduces the induction time of the mixture. Additionally, temperature and composition fluctuations in the unburnt mixture can lead to boundary conditions in the end gas that further promote the occurrence of auto-ignition, as they result in locally accelerated chemical reactions at the exothermic centers[4] that are characterized by a local temperature higher than the mean value [156]. Because the transition to knock does not only depend on the unburnt temperature and pressure, but also on the size, distribution and properties of exothermic centers in the end gas [89] [172], severe knock can occur even at mean unburned gas temperatures that are lower than those required for an ideal homogeneous thermal explosion. Generally, depending on the boundary conditions, an exothermic center can lead to the occurrence of deflagration (low mean end gas temperature and steep temperature gradients resulting in light knock or acting as a flame extension), thermal explosion (high mean end gas temperature and small temperature gradients resulting in moderate knock), or developing detonation (intermediate mean end gas temperature and temperature gradients resulting in severe knock) [89] [172]. However, pure thermal explosions are reported to be unlikely in internal combustion engines. The developing detonation mode is the regime that is most likely to cause serious knock damage [89] [130]. It is

---

[4] Very often referred to as "hot-spots" in the literature [135] [168], although this term is also commonly used in the context of surface ignition, where it describes an overheated valve or spark plug or a glowing deposit on the combustion chamber walls [75].

commonly assumed that auto-ignition occurs at several locations in the end gas quite simultaneously and the three regimes (deflagration, explosion and detonation) cannot be clearly distinguished, as they appear in mixed forms [130]. Investigations have revealed that high turbulence of the engine charge reduces the knock propensity in general, but does not eliminate the temperature fluctuations in the end gas completely [91] [129].

2.1.2 Auto-Ignition in the End Gas

The common fundamental knock theories are based on the description of hydrocarbon oxidation leading to auto-ignition in the end gas zone [75] [88]. This oxidation process is typically illustrated with simple hydrogen-oxygen systems on a molecular level and by using elementary chemical reactions, as for the hydrocarbons commonly found in real gasoline fuels, the chemical reaction pathways by which the fuel molecules are broken down and react to form products are very complicated [75] [150] [158]. These consist of a large number of simultaneous, interdependent chain reactions [62].

The hydrocarbon oxidation process is started by an initiating reaction, where highly reactive intermediate species, referred to as radicals, are produced from the stable fuel and oxygen molecules [85]. This step is followed by propagation reactions, where the produced radicals react with the reactant molecules to form products and other radicals that continue the chain. The process ends with termination reactions, where the chain-propagating radicals are consumed. Some propagating reactions, referred to as chain-branching, produce two reactive radical molecules for each radical consumed [158]. As soon as the number of radicals increases sufficiently rapidly due to chain branching, the reaction rate becomes extremely fast. In this case, the energy released by the chemical reactions in the form of heat is larger than the heat lost to the surroundings [75]. The resulting temperature and pressure increase accelerates the rates of all following chemical reactions, as these have been proven to show exponential temperature dependence [4] [5], thus leading to an auto-ignition of the air-fuel mixture [11] [62].

The boundary conditions in the unburnt mixture, such as temperature, pressure as well as the fuel and mixture composition, are decisive for both the type of chemical reactions that take place during the hydrocarbon oxidation process (initiating, branching, propagating, termination) and the reaction rates [62]

[75] [150]. In the lower temperature range, chain branching reactions slowly build up the radical pool and lead to a decrease of the induction time (mixture ignition delay). At higher temperatures in the intermediate range (negative temperature coefficient zone), the radicals decompose back to their reactants because of their instability, causing a negative temperature dependency of the induction time [158]. In this case, which is also known as "cool flame" occurrence, only a small fraction of the reactants has reacted and the temperature rise resulting from the auto-ignition is only tens of degrees [75]. Depending on the boundary conditions and the fuel, a cool flame might be followed by the occurrence of a second, high-temperature ignition ("hot flame"), thus showing a two-stage ignition behavior [150]. It is important to remark that while all hydrocarbons are characterized by induction intervals, which are followed by a very rapid reaction rate, some hydrocarbon compounds[5] do not exhibit the cool flame or two-stage ignition behavior[62] [75]. Additionally, as the temperature of the mixture increases, a transition from two-stage to single stage ignition takes place [158]. The occurrence of cool flames, two-stage as well as a transition to single stage auto-ignition has also been observed in motored engines [46] [75] [165]. At temperatures higher than the intermediate regime, the fast high-temperature reactions dominate, resulting in typically short mixture ignition delay times.

The reaction pathways and rate rules shall not be further discussed here, as an overview as well as detailed information is available in numerous publications [27] [29] [62] [63] [64] [75] [112] [118] [150] [158] [163] [164].

2.1.3 Knock Detection

Knocking combustion is initiated by an auto-ignition in the unburned mixture and is typically characterized by high-frequency pressure oscillations resulting from pressure waves traveling through the combustion chamber at the speed of sound [75]. Because the combustion chamber walls reflect these waves, the frequency of the pressure oscillations is closely related to the combustion

---

[5] For example fuels with high ethanol content, as ethanol is characterized by single-stage auto-ignition behavior [142] [151].

chamber geometry [124]. Furthermore, the maximum amplitude of the pressure oscillations depends on the boundary conditions at the moment of auto-ignition [108].

There are many possible ways to detect knock in general, e.g. by using vibration sensors for measuring high-frequency vibrations (commonly used today in commercial engines), the trace of the in-cylinder pressure trace measured by a piezoelectric transducer (typically installed on research engine test benches), optical access (research purposes, as very difficult to implement) or acoustic sensors for the detection of audible knock [113]. Without an optical access to the combustion chamber, the knock detection based on the in-cylinder pressure trace is the most reliable method [142]. However, during knocking combustion, the pressure trace recorded by the piezoelectric transducer is no longer representative for the entire combustion chamber [142]. Hence, appropriate definitions and criteria are needed for the detection of knock. In this case, a knock criterion can be based on either the in-cylinder pressure itself or the estimated heat release rate [168], which is strongly related to the measured pressure signal. In this context, it has to be considered that because of the cycle-to-cycle variations, at knock limited spark advance many of the single engine cycles do not contain knocking events. Furthermore, the abnormal combustion events in the knocking single cycles have strongly varying intensities. Hence, it has to be distinguished between the detection of the knock limited spark advance, the differentiation between knocking and non-knocking single cycles as well as the estimation of the cycle-individual knock onset, which is of particular importance for modeling knock. The detection of the knock onset of single cycles will be discussed in the context of preprocessing the measurement data used for the development of the new 0D/1D knock model in Section 3.2.1.

Different characteristic numbers for the knocking combustion (knock indices) can be obtained from the recorded in-cylinder pressure signal [174]. They are all affected by the data filter (typically high-pass or band-pass in the region of $4 - 25$ kHz [142]), the in-cylinder location of the pressure transducer as well as the data sampling resolution [20]. The most common knock indices are the knock peak-to-peak amplitude (KPP) [28], the maximum amplitude of the pressure oscillations (MAPO) [174] and the integral of the modulus of pressure oscillations (IMPO), which considers both the amplitude and the length of the oscillations [168]. Numerous other characteristic numbers have been proposed over the years as well, e.g. the integral of the modulus of the pressure gradient (IMPG), the value of the third derivative of the in-cylinder pressure signal and the signal energy of the pressure oscillations (SEPO). A detailed description of the various knock indices can be found in [168]. No matter which characteristic number is used, it should be independent of the operating point [142]. To this end, the KPP value has to be referred to the engine speed [28]. Additionally, [30] found out that the MAPO and IMPO of the non-knocking cycles increase with engine speed, so that it is recommended to reference these two indices to the non-knocking value.

Finally, as the knock index is calculated for each individual cycle, the number of the recorded knocking single cycles can be used for the definition of a knock boundary (knock limited spark advance) that is characterized by a knock frequency. To this end, first a threshold value for the selected knock index has to be defined, so that it can be determined whether a specific single cycle is knocking or not. Thereafter the knock frequency can be calculated by dividing the number of knocking cycles by the total number of recorded cycles. Lastly, the knock boundary is defined as a knock frequency window or a threshold. This enables the assessment if an operating point (characterized by its spark timing) is beyond (severe knock), at (light / medium knock), or close to (no knock) the knock limited spark advance.

2.2   Kinetic Modeling and Gasoline Surrogates

As already discussed in Section 2.1.2, numerous publications are available on the pathways and rate rules of the hydrocarbon oxidation process leading to auto-ignition in the end gas that precedes the occurrence of knock. Additionally, many reaction kinetic models for different (conventional and non-conventional) gasoline fuels have been formulated in the recent decades. These are nowadays often employed in computational research as well as in the development of internal combustion engines [54]. The kinetic models have different levels of detail and are often a trade-off between computational cost and accuracy. The number of included reaction steps and species are a measure for the model's level of detail.

In the context of modeling engine knock, kinetic mechanisms are commonly used for the simulation of ignition delay times at various boundary conditions. The development of kinetic models is facilitated by the application of reaction classes and rate rules [27]. Precise and versatile rate rules are desirable to improve the prediction quality of kinetic models, which are typically calibrated against measurements of ignition delay times, laminar burning velocities, and stable species profiles at different pressures from flow and jet stirred reactors [28]. Similarly, the accuracy of kinetic models is assessed based on literature data as well as, for the purpose of modeling engine knock, ignition delay times measured with rapid compression machines (RCM) and shock tubes (ST) [26]. As these two appliances have fundamentally different working principles, facility effects of both setups have to be taken into account in the course of the kinetic model development. In this context, the measurement error tolerances of the two facilities differ significantly – typically 10 % for RCM and, in case of a ST, up to 20 % [24].

Moreover, the magnitudes of the measured ignition delay times vary significantly (microseconds to seconds). The ignition delay times in the low-to-intermediate temperature regimes are typically obtained in rapid compression machines. A RCM generally consists of three main components –.a pneumatically driven piston, a hydraulic braking, and a control chamber with a reactor piston [96]. The reactor volume can be varied through the interchange of the end walls. The cylindrical reactor chamber has a handful of accessible ports for gas inlet / outlet valves, optical, mass sampling, and pressure measurements [28]. A RCM is further equipped with an external heating system covering the reactor chamber. The two non-idealities characterizing rapid compression machines are heat loss and radical pool generation during the compression phase, both affecting the mixture reactivity behavior and thus the measured ignition delay time. It is therefore essential to consider these effects in the course of the development and validation of kinetic models [143].

Shock tubes are typically used in the high temperature range, where the ignition delay is very short and cannot be determined in a rapid compression machine [173]. Compared to RCM, shock tubes can increase the temperature and pressure of the mixture very quickly, thus eliminating the effects of the heating or compression processes. In a ST, high-pressure driver gas and the low-pressure air-fuel mixture are separated through a double diaphragm chamber housing up to two aluminum diaphragms. When the diaphragm breaks, a shock wave is generated and is reflected at the air-fuel mixture's side, thus bringing the test gas to high pressure and temperature. The residence time for the elevated pressure is typically low and depends on the facility [142]. However, in the case of relatively long ignition delay times, the development of boundary layers can affect the mixture reactivity behavior [28].

While several global properties of practical fuels, e.g. octane number and energy density, can be either experimentally determined or numerically calculated, detailed combustion kinetics are very complex and therefore highly sensitive to the fuel constituents [26] [28] [53]. Because real petroleum fuels are composed of a huge variety of hydrocarbon components, kinetic models are typically employed in combination with simplified, surrogate hydrocarbon mixtures for the detailed description of the combustion kinetics [54]. For gasoline fuels that are generally used in spark ignition and homogeneous charge compression ignition engines, the mixture of n-heptane and iso-octane, also known as Primary Reference Fuel, is often suggested as an appropriate surrogate [26] [40]. Additionally, in order to account for aromatic compounds in real gasoline fuels, ternary mixtures of n-heptane, iso-octane, and toluene are often proposed [60] [92]. In the case of ethanol-doped gasoline, mixtures of n-heptane, iso-octane, toluene, and ethanol are commonly used for mimicking the target properties of the real fuel [54] [117].

Generally, only one specific surrogate composition is representative for a given real gasoline fuel and matches up to its characteristics [53]. In this context, the values of numerous fuel properties, such as the research and motor octane numbers, hydrogen to carbon ratio and the liquid density are highly important for the accurate description of the real gasoline's auto-ignition behavior [38] [54]. For this reason, the surrogate composition, which is estimated with appropriate blending rules [2] [26] [54] [114] [117], should be additionally optimized by minimizing all property differences between the real and surrogate fuels [28].

2.3   0D/1D Simulation of Internal Combustion Engines

The available models for predicting knock in 0D/1D SI engine simulations are commonly based on the Livengood-Wu integral [105] and generally known for their poor performance and the great effort needed for their calibration [52] [53]. This fact results in significant restrictions on the development of future SI engine concepts within a 0D/1D simulation environment, because a reliable, fully predictive knock model is an essential requirement for accomplishing

this task. In the following sections, the fundamentals of the 0D/1D spark ignition engine simulation as well as the state-of-the-art knock modeling approaches are presented, as these are essential for the development of new, fully
predictive 0D/1D simulation model.

2.3.1 Fundamentals of the 0D/1D Spark Ignition Engine Simulation

The calculation of in-cylinder processes involves both the estimation of the
gas exchange with at least one valve opened (low-pressure phase) and the computation of the combustion with closed valves (high-pressure phase). Generally, two different types of calculations exist, depending on the target: analysis
and simulation. Regarding the high-pressure phase, this means that in the case
of an analysis, a combustion process is calculated from a known pressure profile with the help of simple thermodynamic assumptions. This calculation type
is also commonly known as pressure trace analysis, PTA. The simulation does
exactly the opposite – an available (directly or indirectly) description of a combustion process is used to estimate the cylinder pressure curve, again based on
thermodynamic assumptions. For achieving the goals of this work, both calculation types are crucial.

Performing a pressure trace analysis only requires cylinder pressure profiles
and some basic engine operation data such as the fuel and air mass flows as
well as the external EGR rate measured on a test bench. In case of a combustion simulation however, the question arises, how the heat release rate can be
estimated. The simplest way of doing this is performing a PTA and thus calculating a heat release rate that can afterwards be used for the simulation of
the analyzed operating point. A reasonable application of this method is the
calculation of certain parameters that were not estimated during the course of
the measurement campaign at the test bench. Apart from that, the use of this
heat release rate calculation method is very limited. Hence, the predictive
0D/1D engine simulation requires an appropriate approach for modeling the
spark ignition combustion. To this end, many empirical models describing the
SI combustion process have been developed. They comprise simple mathematical functions contacting a handful of different parameters that have to be
calibrated so that the heat release rates calculated by the empirical model re-
semble those estimated with a PTA.

Wiebe [155] developed the most popular empirical combustion model, which has been successfully utilized by many researchers, e.g. [61]. The model is capable of reproducing the heat release rate shape of SI engines very well [85], however it is not predictable, as it requires the individual recalibration of the model parameters at each engine operating point. Therefore, the original Wiebe approach is also often referred to as a "substitutive" heat release rate. Over the years, many extensions have been developed that employ empirical equations for the calculation of the model parameters as a function of the operating conditions [39] [76] [167]. Thus, model predictability has been partially achieved. However, the prediction capabilities of the empirical SI combustion models remains limited as the models themselves as well as their extensions, if any, have a purely empirical background [85]. For this reason, this type of combustion models is not of interest for the purposes of this work and will not be further discussed.

Phenomenological combustion models on the other hand generally have exceptional prediction capabilities, as they use correlations that are consistent with the fundamental theory. Based on assumptions and simplifications, they describe the physical and chemical effects relevant to the combustion progress directly. Thus, full model predictability can be achieved for the combustion process of interest. Ideally, the model calibration at a single measured operating point is sufficient for the reliable and accurate simulation of the entire engine map. The broad use of this model class, for example in [7] [67] [135] [162], is also associated with the low computation times, which are typically within the range of milliseconds to a couple of seconds. At the same time, the achieved quality of the calculation results is generally high.

Finally, it should be remarked that an even more detailed description of the processes in the combustion chamber can be achieved by using 3D CFD simulation models, where the combustion chamber is divided into a large number of cells (discretization). Subsequently, all state variables for each cell are calculated by taking the conservation equations and other laws into account. However, in contrast to the phenomenological models, the 3D CFD computation times for a single engine cycle are in the range of days rather than seconds [35]. Hence, such modes typically have a different application field than engine concept design, for example, the detailed optimization of geometries or single components.

Phenomenological combustion modeling constitutes the foundation of the 0D/1D knock simulation and is discussed in detail in Section 2.3.1.2. Furthermore, as averaging a signal generally leads to an information loss, not all data present in individual engine working cycles is contained in the corresponding representative average cycle. Hence, the simulation of single engine cycles and thus the cycle-to-cycle variations, which are typical for SI engines [75], seems to be a promising approach for improving the quality of the knock boundary simulation, as it reproduces the pressure and temperature curves of the single cycles. For this reason, a phenomenological model for the simulation of cycle-to-cycle variations [162] that can be used to evaluate the influence of the combustion fluctuations on the knock simulation is presented in Section 2.3.1.3. However, as the laws of thermodynamics are fundamental for the description of the processes in an internal combustion engine, they shall be briefly reviewed in the first place. A deeper insight into the fundamentals of thermodynamics is available in [66] [75] [122].

2.3.1.1 Basic Concepts of Thermodynamics

Thermodynamics is concerned with the mathematical modeling of the real world. The starting point for describing the processes in an internal combustion engine is the modeling of the combustion chamber as a thermodynamic system. A thermodynamic system is a quantity of matter of fixed identity, around which a boundary can be drawn. Everything outside the boundary is referred to as surroundings. The system boundary can be fixed or moveable and must be clearly defined. Furthermore, it can be distinguished between open systems, in which matter crosses the system boundary, and closed ones. Open systems can exchange both matter and energy with the surroundings, whereas in closed systems only an energy exchange is possible, usually in the form of work or heat. Sometimes instead of system, the term control volume is used. In the case of a closed system, in which the mass of matter remains constant, the control volume is referred to as control mass. A control volume is enclosed by a control surface.

The thermodynamic state of a system is defined by specifying the values of a set of measurable properties, sufficient to determine all other system properties. For fluid systems, typically pressure, volume, and temperature are used. A change in the state of the thermodynamic system is called a process. When the initial and final states of a process are the same, the process is referred to as cycle. Figure 2.1 shows the thermodynamic system "combustion chamber" and its boundary.

A system can further be divided into different zones (sub-systems) that must not overlap. If the properties defining the thermodynamic state of a system are extensive (e.g. volume), the value of the property for the whole system equals the sum of the values for the different sub- systems. Intensive properties (e.g. temperature and pressure) on the other hand do not depend on the quantity of matter present. The value of the extensive properties per unit mass yields the specific properties of a system. They are intensive, because they do not depend on the mass of the system.

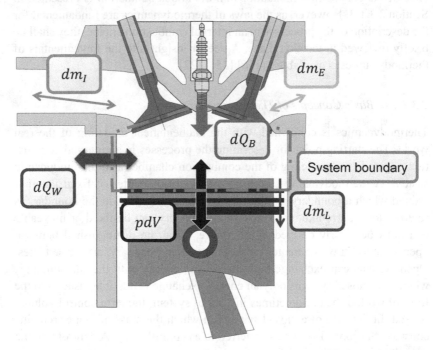

Figure 2.2:   The combustion chamber as a thermodynamic system [66] [85].

In the context of modeling combustion engines, it can be assumed that the pressure within a thermodynamic system is locally constant and time-dependent. Furthermore, the system is considered to contain only gaseous components [66]. Hence, all system zones have the same pressure, but may differ in their temperature and gas composition. Additionally, the temperature and the components are considered to be evenly distributed within each thermodynamic sub-system, resulting in temperature and gas homogeneity in each zone.

The most common ways of dividing the combustion chamber into sub-systems are the assumption of complete homogeneity throughout the combustion chamber (single zone, no sub-systems) as well as the two-zone modeling approach, Section 2.3.1.2, where a burnt and an unburnt zone are distinguished during the combustion. For each zone of the system "combustion chamber" the principle of conservation of energy (first law of thermodynamics), Equation 2.1, as well as mass, Equation 2.2, have to be always satisfied. This is also true for the ideal gas law, Equation 2.3.

$$\frac{dQ_B}{d\varphi} + \frac{dQ_W}{d\varphi} + p_{cyl} \cdot \frac{dV}{d\varphi} + h_E \cdot \frac{dm_E}{d\varphi} + h_I \cdot \frac{dm_I}{d\varphi} + h_L \cdot \frac{dm_L}{d\varphi} = \frac{dU}{d\varphi} \quad \text{Eq. 2.1}$$

φ	crank angle [°CA]
$\dfrac{dQ_B}{d\varphi}$	heat release rate [J/°CA]
$\dfrac{dQ_W}{d\varphi}$	wall heat flux [J/°CA]
$\dfrac{dV}{d\varphi}$	volume change [m³/°CA]
$\dfrac{dm_E}{d\varphi}$	exhaust mass flow [kg/°CA]
$\dfrac{dm_I}{d\varphi}$	inlet mass flow [kg/°CA]
$\dfrac{dm_L}{d\varphi}$	leakage mass flow (blowby) [kg/°CA]
$\dfrac{dU}{d\varphi}$	internal energy change [J/°CA]
p_{cyl}	cylinder pressure [Pa]
h_E	specific exhaust enthalpy [J/kg]
h_I	specific inlet enthalpy [J/kg]
h_L	specific leakage enthalpy [J/kg]

During the combustion, the valves are (generally) closed and the fuel (injection) as well as leakage (commonly referred to as engine blowby [75]) mass flows can usually be neglected. Thus, there are no mass flows to consider and the equations for the calculation of the high-pressure phase are even simpler than the general equation forms presented in this section. In this case, if all gas properties (internal energy, enthalpy, and specific gas constant) and the wall heat losses are known, the only missing parameter in Equation 2.1 needed for the calculation of the cylinder pressure profile is the heat release rate, which can be estimated as discussed in Section 2.3.1.2.

$$\frac{dm}{d\varphi} = \frac{dm_I}{d\varphi} + \frac{dm_E}{d\varphi} + \frac{dm_L}{d\varphi} + \frac{dm_F}{d\varphi} \qquad \text{Eq. 2.2}$$

$\dfrac{dm}{d\varphi}$ total mass flow / cylinder mass change [kg/°CA]

$\dfrac{dm_F}{d\varphi}$ flow of injected fuel mass [kg/°CA]

$$p_{cyl} \cdot \frac{dV}{d\varphi} + V \cdot \frac{dp_{cyl}}{d\varphi} = m \cdot R \cdot \frac{dT}{d\varphi} + m \cdot T \cdot \frac{dR}{d\varphi} + R \cdot T \cdot \frac{dm}{d\varphi} \qquad \text{Eq. 2.3}$$

$\dfrac{dp_{cyl}}{d\varphi}$ cylinder pressure change [bar/°CA]

$\dfrac{dT}{d\varphi}$ temperature change [K/°CA]

$\dfrac{dR}{d\varphi}$ individual gas constant change [J/kg/K/°CA]

V volume [m³]

m mass [kg]

R individual gas constant [J/kg/K]

T temperature [K]

A handful of approaches for the calculation of the wall heat losses exist [8] [10] [66] [169]. The estimation of the gas properties is presented in detail in

[65] and [68]. Additionally, [65] and [66] discuss methods for solving the differential system given by Equation 2.1, Equation 2.2 and Equation 2.3. The next section presents the phenomenological two-zone combustion model used for the estimation of the still needed heat release. All thermodynamic calculations introduced in this section are integrated in and can be performed with the FVV cylinder module [66] – a powerful open-source software tool that was used for accomplishing the tasks of this work.

2.3.1.2   Two-Zone Combustion Modeling

A commonly used approach for modeling the combustion in conventional spark ignition engines is the Entrainment model [66] [67] [146]. The model assumes a hemispherical flame starting at the spark plug and propagating with a speed always perpendicular to its surface, Figure 2.3.

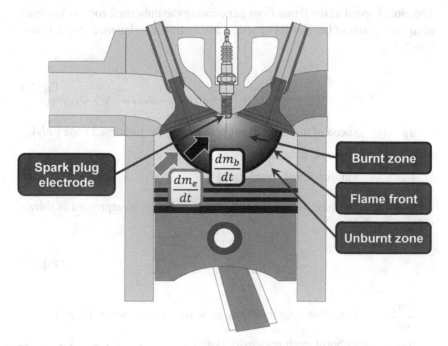

Figure 2.3:   Schematic representation of the Entrainment model [85] [161].

For the calculation of the flame surface, the spark plug position in relation to the cylinder walls is needed. In case of a central spark plug position, the resulting flame surface would increase extremely at first, until it reaches the walls, and then diminish. This effect would lead to very unrealistic heat release rates with a very sharp peak. For this reason, the spark plug should not be set exactly at the center of the combustion chamber. A slight spark plug position offset on the other hand leads to an unsymmetrical flame geometry, thus correctly representing the deviations from a perfectly spherical propagation that are always observed in reality.

The combustion chamber is divided into three sub-systems: unburnt zone, burnt zone and flame front, which separates these two zones. However, the exact properties of the flame are not calculated – it is assumed that the flame is a part of the unburnt zone, resulting in a two-zone modeling approach.

The global speed of the flame front penetrating the unburned zone is assumed to equal the sum of laminar flame speed and isotropic turbulence speed, Equation 2.4.

$$u_E = u_{Turb} + s_L \qquad \text{Eq. 2.4}$$

u_E	speed of the flame front penetrating the unburned zone [m/s]
u_{Turb}	isotropic turbulence speed [m/s]
s_L	laminar flame speed [m/s]

Consequently, the mass brought into the flame zone can be expressed as shown in Equation 2.5.

$$\frac{dm_E}{dt} = \rho_{ub} \cdot A_{Fl} \cdot u_E \qquad \text{Eq. 2.5}$$

$\dfrac{dm_E}{dt}$	mass flow into the flame zone (mass entrainment) [kg/s]
ρ_{ub}	unburnt mixture density [kg/m^3]
A_{Fl}	flame surface [m^2]
t	time [s]

The entrained mass, together with the mass flow into the burnt zone, can be combined in a differential equation for the estimation of the change of the flame zone mass. With the additional definition of a characteristic burn-up time, the desired heat release rate is yielded by Equation 2.6. The characteristic burn-up time itself, Equation 2.7, describes the time needed for the laminar combustion of a turbulence eddy with the size of the Taylor length[6]. The Taylor length is generally used for characterizing turbulent flows and can be estimated from the integral length scale, turbulent velocity and turbulent kinematic viscosity, Equation 2.8. The Taylor length coefficient needed for its calculation is set to 15, as proposed by Heywood [75]. The integral length scale is a measure of the size of the large energy-containing structures of the flow [49] [75] and can be assumed to equal the diameter of a sphere with the combustion chamber volume [85].

$$\frac{dm_b}{dt} = -\frac{dm_{ub}}{dt} = \frac{dQ_B}{d\varphi} \cdot \frac{1}{H_u} \cdot \frac{d\varphi}{dt} = \frac{m_F}{\tau_L} \qquad \text{Eq. 2.6}$$

$\dfrac{dm_b}{dt}$ mass flow into the burnt zone [kg/s]

$\dfrac{dm_{ub}}{dt}$ mass flow into the unburnt burnt zone [kg/s]

$\dfrac{d\varphi}{dt}$ change of crank angle over time [°CA/s]

H_u lower heating value [J/kg]

m_F flame zone mass [kg]

τ_L characteristic burn-up time [s]

$$\tau_L = \frac{l_T}{s_L} \qquad \text{Eq. 2.7}$$

l_T Taylor length [m]

---

[6] Also known as Taylor microscale and Taylor micro length scale.

Thus, only the laminar flame and the isotropic turbulence speeds are yet unknown. A handful of empirical correlations for the calculation of laminar flame speeds at different boundary conditions are available. The two-zone combustion model incorporates a slightly modified version [65] of the correlation originally proposed in [75] that is shown in Equation 2.9. It should be remarked that recently, new laminar flame speed models have been developed based on kinetic simulations with detailed kinetic reaction mechanisms, see Section 4.1. These account for the effects of various parameters and are supposed to yield much more accurate results over a significantly wider boundary condition range [71].

$$l_T = \sqrt{\chi_T \cdot \frac{v_{Turb} \cdot l}{u_{Turb}}}$$

Eq. 2.8

χ_T Taylor length coefficient [-]

v_{Turb} kinematic turbulent viscosity [m²/s]

l integral length scale [m]

$$s_L = \left(0.305 - 0.549 \cdot \left(\frac{1}{\lambda} - 1.21\right)^2\right) \cdot \left(\frac{T_{ub}}{298\,K}\right)^{2.18-0.8\cdot\left(\frac{1}{\lambda}-1\right)}$$
$$\cdot \left(\frac{p}{10^5\,Pa}\right)^{-0.16+0.22\cdot\left(\frac{1}{\lambda}-1\right)} \cdot \left(1 - 2.06 \cdot x_{AGR,st}{}^{\xi}\right)$$

Eq. 2.9

T_{ub} unburnt mixture temperature [K]

λ air-fuel equivalence ratio [-]

$x_{EGR,st}$ stoichiometric exhaust gas recirculation rate[7] [-]

ξ exponent for the influence of exhaust gas[8] [-]

---

[7] Definition proposed in [90] to account for the unburnt air mass in the exhaust gas in case of a lean combustion.

[8] Heywood [75] proposed a value of 0.77. However, in the model presented here the exponent for the exhaust gas influence on laminar flame speed is set to 0.973 [67].

The isotropic turbulence speed is a function of the specific turbulence in the combustion chamber, Equation 2.10, which is estimated with a separate turbulence model. The most popular phenomenological approach for modeling the specific turbulence are the k-ε turbulence models, which estimate the changes in the turbulence level by evaluating various production and dissipation terms [8] [17] [85] [90] [120] [161].

$$u_{Turb} = \sqrt{\frac{2}{3} \cdot k}$$
Eq. 2.10

k specific turbulence $[m^2/s^2]$

Finally, the level and thus influence of turbulence on the combustion is set by adjusting the parameter C_k[9], which scales the starting value of the specific turbulence [67]. This is the only calibration parameter of the combustion model. Its value is engine-specific and its accurate estimation at a single measured operating point is sufficient for the reliable and accurate simulation of the entire engine map.

2.3.1.3 Phenomenological Modeling of Cycle-to-Cycle Variations

The cycle-to-cycle variations (CCV) model proposed in [162] is capable of predicting the cycle-to-cycle variations over the entire engine map after a calibration of two main parameters at a few measured operating points. It considers the influences of engine speed, load, air-fuel equivalence ratio, residual gas content and to a certain extent the effects of the variation of valve timing and intake valve lifts. The CCV model implementation is based on the assumption that the cyclic fluctuations of combustion can be modeled by a variation of the parameters available in the combustion model (e.g. the turbulence level, Sec-

---

[9] If the turbulence model proposed in [17] is used, the turbulence level's influence on combustion is defined by the parameter C_u, which slightly differs from C_k. In this case, the turbulence model phenomenologically estimates the starting value of the specific turbulence and the adjustment of C_u scales the isotropic turbulence speed in Equations 2.4 and 2.8.

tion 2.3.1.2). Starting from the combustion model calibration for the simulation of the mean cycle, an alternation of the values of selected model parameters is performed.

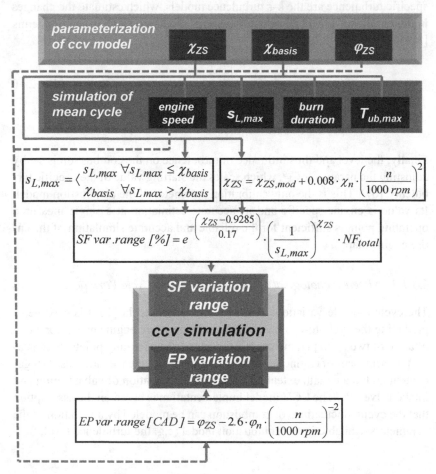

Figure 2.4:   Overview of the phenomenological CCV model [162].

The model assumes that two main factors influence the cyclic fluctuations – the charge dilution and the combustion position. These are modeled by means of two separate parameter variations. Thus, the CCV model has two main calibration parameters (χ_{ZS} and φ_{ZS}) that are used for the calculation of the combustion model parameter variation ranges, as shown in Figure 2.4. The calibration parameters can be calibrated independently in different engine map regions, as they have different ranges of action (part load and full load), making the model very user-friendly.

A third, supplementary parameter χ_{basis} affects the intermediate operating range, where the ignition timing and thus the center of combustion (represented by 50 % mass fraction burnt point, MFB50) are optimal and the charge dilution is at a minimum degree. However, the cyclic variability level in this operating range is small compared to the CCV level at full load or at lower part load, where efficiency-boosting EGR strategies are usually applied. Thus, the supplementary parameter χ_{basis} is of minor importance and generally does not have to be calibrated. The model commonly estimates 15 single cycle per operating point, with this value being adjustable. However, in this context it should be kept in mind that the total computational time is proportional to the number of simulated single cycles.

2.3.2 0D/1D Knock Modeling Fundamentals

Many researchers have attempted to model knock using various approaches. As knock occurrence results from an auto-ignition and thus chemical reactions, one of the main challenges is to find an appropriate and accurate description for the real in-cylinder chemical processes, which can be incorporated in a 0D/1D simulation environment. The common approaches used for modeling knock can be classified into three categories:

- Detailed reaction kinetic mechanisms

- Reduced reaction kinetic mechanisms

- Phenomenological approaches

Overall, a lot of effort has been put into the development of kinetic mechanisms for gasoline fuels in the recent years. Nowadays it is possible to employ these for the prediction of knock in the course of 3D CFD simulations [14] [101] [102] [104] [123] [157] [175]. This knock modeling approach enables the evaluation of local thermodynamic states and velocities as well as species concentrations and thus the prediction of the exact locations of knock onset.

However, knock is a stochastic phenomenon that strongly depends on the cycle-to-cycle variations and the experimental knock limited spark advance is characterized by a knock frequency, Section 2.1. Thus, the exact representation of this phenomenon in 3D CFD requires the large eddy simulation (LES) of a large number of engine cycles. Furthermore, detailed mechanisms for gasoline fuels are comprised of hundreds of species and thousands of elementary reactions, Section 2.2. Thus, LES simulations combined with detailed mechanisms are characterized by an overall huge computational cost. Therefore, this knock modeling approach is typically used for research purposes [28] and is not suitable for the investigation of multiple engine configurations and operating condition variations. Alternatively, the cycle-to-cycle variability can be mimicked within the Reynolds-Averaged Navier–Stokes (RANS) framework by adjusting the initial flame kernel in the combustion simulation resulting in different combustion phasing while maintaining constant spark timing [28] [59]. Additionally, many methods for the automatic kinetic mechanism reduction aiming at optimizing the computational performance whilst preserving the result accuracy have been developed. However, even with reduced mechanisms, the mesh size and the simulation time step in both RANS and especially LES simulations still need to be sufficiently small, which again results in long computational times.

Alternatively, reaction kinetic mechanisms, detailed or reduced, can be coupled with quasi-dimensional combustion [19] [115] or stochastic reactor [117] models. However, this knock modeling approach shall not be discussed in detail here because the computational times are by magnitudes longer than the values typical for 0D/1D engine simulations (a couple of seconds per cycle). Thus, the simulation of knock with detailed or reduced kinetic mechanisms is of limited suitability for the investigation of multiple engine configurations and operating condition variations within a 0D/1D environment.

Phenomenological approaches on the other hand use empirical correlations that are consistent with the fundamental theory describing the real in-cylinder chemical processes. Thus, they are very simple and yet have high prediction capabilities [140]. For these reasons, most 0D/1D knock models that have been developed in the past 50 years have a phenomenological background. In this context, one "simplified chemistry" approach stands out and underlies the majority of the 0D/1D knock models commonly used today [34] [140]. It involves the evaluation of an integral representing a single global chemical reaction resulting in auto-ignition in the unburnt mixture. The following section presents in detail this phenomenological approach, commonly known as the Livengood-Wu / knock integral.

2.3.2.1 Livengood-Wu Integral

The prediction of auto-ignition of an air-fuel mixture in an internal combustion engine is a challenging task. The mixture has a state-time history that is changing, as it is being compressed and expanded in the engine cylinder. To this end, in 1955 Livengood and Wu proposed a correlation that makes the prediction of the ignition delay time of an air-fuel mixture in motored and firing spark ignition engines possible, Equation 2.11 [105].

$$1 = \int_{t=0}^{t=t_e} \frac{1}{\tau} dt \qquad \text{Eq. 2.11}$$

t elapsed time [s]

t_e time at integration end / overall auto-ignition reaction time [s]

τ mixture ignition delay at the current boundary conditions [s]

Equation 2.11 implies that, in order to predict when auto-ignition of a mixture in a firing engine will occur, the ignition delay times at the corresponding boundary conditions have to be known at each integration step. Their values describe the time interval required for the chemical reactions to result in auto-ignition of the air-fuel mixture, assuming that the physical state of the mixture is constant. Hence, the integral correlation relies on the ignition delay under constant state as an indicator of reactivity. In general, the ignition delay times can be measured in a rapid compression machine or a shock tube, see Section 2.2. Subsequently, an Arrhenius-type equation [4] [5] can be calibrated by estimating the empirical constants C, C_1 and C_2 and then used for their calculation, Equation 2.12. It should be remarked that, **for commonly used gasoline fuels, Equation 2.12 is only adequate in a limited range of temperature for any particular pressure** [15] [136]. Although this has been known for more than 50 years, 0D/1D knock models still typically incorporate modified versions of this equation, as discussed in Section 2.3.2.2.

$$\tau = C \cdot p^{C_1} \cdot e^{C_2/T} \qquad\qquad \text{Eq. 2.12}$$

τ ignition delay of the air-fuel mixture [s]

p current pressure [Pa]

T current temperature [K]

C, C_1, C_2 empirical constants [$1/\text{Pa}^{C_1}$, -, K]

Generally, the auto-ignition of an air-fuel mixture is preceded by and results from many chemical reactions that take place. The Livengood-Wu integral in Equation 2.11 is a simplified representation of the degree of progress of these chemical reactions. It is based on a couple of assumptions with the first one being that an equivalent aggregate reaction with gross parameters as shown in Equation 2.13 exists and describes the progress of the auto-ignition process [107]. Hence, there is a fixed functional relationship between time, the rate of this reaction and the instantaneous physical state of the mixture. The reaction rate is not constant during the ignition delay period [105]. It is further assumed that the reactions in a rapid compression machine, motored and firing engines are of the same kind [43] [47], as also discussed in Section 4.4.1. The assumed aggregate reaction results in a sudden transition to a process that completes the combustion reaction. This happens in a time interval that is much smaller than the preceding delay period [105]. Thus, the occurrence of high-speed chemical reactions marks the end of the delay period and the ignition duration can be neglected when compared to the ignition delay.

$$\frac{d(x)}{dt} = \phi_1(p, T, t), \phi_2(F, chemical\ composition, etc.) \qquad \text{Eq. 2.13}$$

x concentration of pertinent reaction products [mol/m³]

t time [s]

p absolute pressure [Pa]

T absolute temperature [K]

F fuel-air ratio [-]

ϕ_1, ϕ_2 empirical functions

This assumed auto-ignition behavior yields the concept of a critical value for the concentration of chain carriers in the mixture $(x)_c$ which, when exceeded, results in auto-ignition. The functional relationship between the chain carrier fraction $(x)/(x)_c$ and the relative time t/τ describes the assumed aggregate reaction leading to auto-ignition, as shown in Equation 2.14 and Equation 2.15. This aggregate reaction represents an integration process during which the concentration of chain carriers builds up with a rate dependent on the current boundary conditions, until the upper integration limit, defined by $(x)_c$, has been reached.

$$\frac{d}{dt}\left[\frac{(x)}{(x)_c}\right] = \phi\left(\frac{t}{\tau}\right) \qquad \text{Eq. 2.14}$$

$$\frac{(x)}{(x)_c} = \int_{t=0}^{t=t_e} \phi\left(\frac{t}{\tau}\right) dt = 1.0 \qquad \text{Eq. 2.15}$$

Moreover, with the assumption of a constant reaction rate during a fixed state process (zero-order reaction) Equation 2.15 can be further simplified, resulting in Equation 2.16 and thus the Livengood-Wu integral as shown in Equation 2.11. In contrast to the assumed zero-order reaction, in reality fuel oxidation processes invariably consist of a network of elementary reactions that can be propagating, branching or terminating, with the associated elementary reaction orders that can be first, second or third, which individually exert either promoting or retarding influences on the net progress of the reactions. However, [119] successfully reformulated the Livengood-Wu integral for a global, n^{th} order reaction and demonstrated that the result degenerates to the integral in Equation 2.16 by interpreting it on an averaged basis. Thus, after decades of quite successful use of the Livengood-Wu integral for the prediction of knock in internal combustion engines, its validity for realistic fuels could be proved.

$$\frac{(x)}{(x)_c} = \int_{t=0}^{t=t_e} \frac{1}{\tau} dt = 1.0 \qquad \text{Eq. 2.16}$$

The value of the Livengood-Wu integral at any given time is also known as pre-reaction state of the mixture. The assumed zero order reaction requires a constant speed of the chemical reactions during every single integration step. As the air-fuel mixture is being compressed and expanded in an engine, this requirement results in a very small (theoretically infinitesimal) integration step. Most importantly, **the critical concentration $(x)_c$ and thus the end-of-integration value of 1 in Equation 2.11 are a characteristic of the aggregate chemical reaction leading to auto-ignition and as such, they are independent of changes in the engine operating conditions**, e.g. increase of engine speed.

As already mentioned, the basic approach discussed in this section has been employed for predicting the auto-ignition delay of air-fuel mixtures in an internal combustion engines for decades. It underlies all commonly used 0D/1D knock models, as discussed in Section 2.3.2.2. However, the assumption of an aggregate global reaction that expresses the chemistry behavior can be interpreted as a "black box" approach [150]. Although the Livengood-Wu integral is very useful for the purposes of knock prediction in 0D/1D simulations, it does not provide a basis for understanding what is actually happening chemically in the unburnt mixture. In this context, in the course of their pioneer research in 1955, Livengood and Wu noted that **interfering effects might arise and impair the prediction quality of Equation 2.11**. For the case of a low-temperature heat release occurring before the auto-ignition for instance, they proposed a separate integration for each ignition stage in succession [105]. Nevertheless, none of the 0D/1D knock models commonly used today accounts for the possible occurrence of two-stage ignition in the unburnt mixture.

2.3.2.2 Review of Phenomenological Knock Models

The majority of the 0D/1D knock modeling approaches commonly used today evaluate the change in the pre-reaction state I_k of an air-fuel mixture and thus the value of a Livengood-Wu integral, as shown in Equation 2.17 [53].

$$I_k = \frac{1}{n} \int_{\alpha = \alpha_{IVC}/90°CA\,bFTDC}^{\alpha = \alpha_{KO}/const.MFB} \frac{1}{\tau} d\alpha \qquad\qquad \text{Eq. 2.17}$$

I_k pre-reaction state of the air-fuel mixture [-]

n engine speed [min^{-1}]

α crank angle [°CA]

The integration is commonly started at inlet valve close (IVC) or 90 °CA before firing top dead center (FTDC) [52]. The calculation ends as soon as the critical concentration of chain carriers $(x)_c$ has been reached and exceeded, resulting in auto-ignition. To this end, a constant critical value for the mixture's pre-reaction state $I_{k,crit}$, which corresponds to $(x)_c$, has to be pre-defined. This end-of-integration point also represents the predicted knock onset (KO). It is further assumed that an auto-ignition after a specific constant mass-fraction-burnt point, usually between MFB75 and MFB85, does not result in knock because of the small unburnt mass and volume fractions left [51] [54] [85]. Additionally, as it has a relatively low temperature, the unburnt mass in the top land is believed to suppress the knock occurrence after flowing out, which happens after the point of maximum cylinder pressure and thus in the late combustion phase [10] [52] [135] [161]. The assumption of a constant latest possible MFB-point where knock can occur results from the fact that many researchers have not observed knock onset after MFB75 / MFB85 [57] [135] [168]. However, their conclusions are based on the investigation of averaged working cycles, thus being affected by the presence of single cycles that do not knock. On the other hand, some knock models postulate that an auto-ignition after MFB95 can also result in knock [45] [69].

Hence, the integration process in Equation 2.17 can also end by reaching the specified constant MFB-point, without having exceeded the pre-defined critical pre-reaction state $I_{k,crit}$. As in this case the mixture does not auto-ignite, no knock occurs. An exemplary evaluation of the knock integral for three single engine cycles is shown in Figure 2.5. Obviously, only one of the plotted cycles is a knocking one, as the other two cycles do not reach $I_{k,crit}$ before MFB85.

With these assumptions, a simple knock controller adjusting the spark timing of the combustion model can be implemented for the simulation of the knock boundary (knock limited spark advance, KLSA). The pre-reaction state curve is always rising, Figure 2.5, as the ignition delay times and thus the integrand values in Equation 2.17 are always positive. After the end-of-integration point, no knock can occur. On the other hand, attaining to the critical pre-reaction state too quickly means that at the latest possible MFB-point where knock can occur I_k will be well above the pre-defined critical value $I_{k,crit}$ and the simulated operating point will be beyond KLSA. Hence, reaching the critical pre-reaction state exactly at the latest possible knocking MFB-point represents the knock boundary in the simulation, which defines the task of the knock controller.

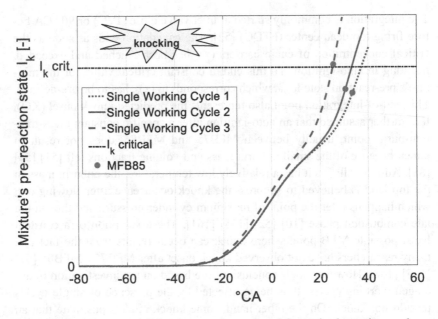

Figure 2.5:   Progress of the pre-reaction state of three single engine cycles.

A detailed look at the knock integral equation unveils the well-known trade-off regarding knock occurrence in SI engines between slow combustion (a lot of time available for auto-ignition to occur) and short burn durations resulting in boundary conditions in the unburnt mixture that promote a faster auto-ignition expressed by higher peak unburnt temperatures and pressures. In the first case, the long burn duration results in a higher number of available integrand values $1/\tau$ that can contribute to the final knock integral value before the combustion ends and a generally longer integration process. On the other hand, fast combustion yields a small number of available integrands; however, these have significantly higher values, as the ignition delay times generally decrease with temperature and pressure, thus leading to a faster knock integral increase.

The only difference between the original Livengood-Wo correlation, Equation 2.11, and Equation 2.17 is the integration domain. When performing numerical integrations over crank angle, it has to be taken into account that changes in engine speed result in different time domain integration steps. Hence, the integration step size in °CA has to be selected small enough, otherwise it will cause problems because of the assumption of constant speed of the chemical reactions during each integration step in the time domain [52]. This will result in errors in the calculated pre-reaction states over engine speed. However, a smaller step size also means longer computational times.

Many variations of the generally used knock prediction approach in Equation 2.17 exist. What all models have in common is that:

- They assume similar simplifications of the physical processes. Generally, the mean values of the unburnt mixture parameters (temperature, pressure etc.) estimated with a phenomenological two-zone SI combustion model, Section 2.3.1.2, are used as inputs for the knock prediction [51] [53]. Furthermore, as already discussed, the integration process always ends at a specified constant MFB-point, if the critical pre-reaction state has not been reached.

- They assume similar simplifications of the chemistry. All models are based on the evaluation of the knock integral. As in the original Livengood-Wu correlation, Equation 2.11, the ignition delay times needed for the integration process are calculated with an Arrhenius-type approach similar to 2.12.

However, despite these basic similarities, the various models available differ considerably in some respects:

- The models generally assume different latest possible MFB-points where knock can occur (end-of-integration), usually between MFB75 and MFB85.

- The knock model inputs are often modified empirically in order to achieve higher knock prediction accuracy. Such expansions aim at representing the real boundary conditions in the unburnt mixture more accurately, e.g. by empirically accounting for mixture and temperature inhomogeneities. Alternatively, the unburnt mixture is sometimes subdivided into a handful of zones that have different temperatures (unburnt zone discretization).

- In some cases, even the base approach, Equation 2.17, is modified and expanded. The added coefficients usually account for turbulence effects that are supposed to influence the knock occurrence.

■ Most importantly, the approaches for the calculation of ignition delay times differ significantly. The main reason for this is the fact that the basic Arrhenius-type Equation 2.12 only accounts for the influences of temperature and pressure. However, many other parameters do not only influence the thermodynamic processes, but also have chemical effects affecting the auto-ignition behavior of the mixture and thus are important for knock occurrence. By adding corresponding terms to Equation 2.12, the influences of such parameters on the ignition delay of the air-fuel mixture can be considered, for example the air-fuel equivalence ratio (AFR) and exhaust gas fraction and thus the EGR rate. Some models also account the influence of the fuel properties represented by the research and / or motor octane number(s). Generally, the estimation of the empirical constants is performed by fitting the knock integral and hence the underlying ignition delay equation to measurement data from an engine test bench that includes variations of the desired parameters by using the least squares method. Alternatively, detailed kinetic reaction mechanisms for gasoline fuels can be used to calculate the ignition delay times of air-fuel mixtures at various boundary conditions and hence to estimate the values of the empirical constants and to expand or modify Equation 2.12, as presented in Chapter 4.

Table 2.1: Different models for knock boundary prediction. The base equation of all models is given by Equation 2.17.

Reference	Equation for the ignition delay τ	Ignition delay equation coefficients			
		C	C_1	C_2	C_3
Douaud et al. [45]	$C\left(\frac{RON}{100}\right)^{C_1} p^{-C_2} e^{\frac{C_3}{T}}$	17.68	3.402	1.7	3800
Frazke [57]	$C\left(\frac{RON}{100}\right)^{C_1} p^{-C_2} e^{\frac{C_3}{T}}$	1	0	1.5	14000
Wayne et al. [159]	$C\left(\frac{RON}{100}\right)^{C_1} p^{-C_2} e^{\frac{C_3}{T}}$	0.389	7.202	1.15	5200
Worret et al. [168]	$C\left(\frac{RON}{100}\right)^{C_1} p^{-C_2} e^{\frac{C_3}{T}}$	0.0187	3.402	1.7	3800
Elmqvist et al. [50]	$C\left(\frac{RON}{100}\right)^{C_1} p^{-C_2} e^{\frac{C_3}{T}}$	0.021	0	1.7	3800
Burluka et al. [25]	$C\left(\frac{RON}{100}\right)^{C_1} p^{-C_2} e^{\frac{C_3}{T}}$	0.01869	3.4017	1.7	3800
Fuiorescu et al. [58]	$(-Cln(n) + C_1)p^{-C_2}e^{\frac{C_3}{TP}}$	8.3	77	2.1	5785

Reference	Equation for the ignition delay τ	Ignition delay equation coefficients			
		C	C_1	C_2	C_3
Schmid et al. [135][10]	$C\left(\dfrac{RON}{100}\right)^{C_1} p^{-C_2} e^{\frac{C_3}{T_P}}$	1.493	0	1.1	4000
Hoepke et al. [77]	$C(1-x_{EGR})^{C_1}\left(\dfrac{p}{T}\right)^{-C_2} e^{\frac{C_3}{T}}$	8.449 x 10^{-5}	0.8881	1.343	5266

		C	C_1	C_2	C_3	C_4
Vancoillie et al. [154]	$C\phi^{C_1}(1+f)^{C_2} p^{-C_3} e^{\frac{C_4}{T}}$	-19.3, T>1050K -22.51, T≤1050K	-0.5559, T>1050K -0.6694, T≤1050K	Polynomial function of T and f	Polynomial function of φ	Polynomial function of φ and f
Chen et al. [34]	$C\left(\dfrac{p}{T}\right)^{-C_1}(1-x_{EGR})^{-C_2} \cdot \lambda^{-C_3} e^{\frac{C_4}{T}}$	5.35 x 10^{-5}	2.374	3.013	1.927	3167

Reference	Equation for the ignition delay τ	i	C	C_1	C_2	C_3	C_4	C_5
Kinetics-Fit [69] [139][11]	$\dfrac{\tau_3(\tau_1+\tau_2)}{\tau_1+\tau_2+\tau_3}$ $\tau_i = C_i\left(\dfrac{RON}{100}\right)^{C_{1,i}} \cdot (fuel)^{C_{2,i}}(O_2)^{C_{3,i}} \cdot (diluent)^{C_{4,i}} e^{\frac{C_{5,i}}{T}}$	1	4.44 x 10^{-7}	3.613	-0.64	-0.564	0.397	12920
		2	11941.423	3.613	-0.64	-1.459	0.486	-1957
		3	8.905 x 10^{-5}	0	-0.25	-0.547	0	16856
Syed et al. [144][12]	$C(1-x_d)^{C_1} p^{-C_2} e^{\frac{C_3}{T}}$	Table						

Table 2.1 summarizes some of the numerous knock modeling approaches available today. The tabulated models propose different approaches for the estimation of ignition delay times; however, the base model in Equation 2.17 has not been modified at all or only slightly. The most popular models are by far those developed by Franzke [57] and Worret [168], the approach based on the work of Douaud and Eyzat in [45] (presumably because it was the first to include coefficients for the fuel research octane number) as well as the Kinetics-

---

[10] T_p is the temperature in a pre-reaction zone in the unburnt mixture, which is being influenced by temperature inhomogeneities (hot-spots) and turbulence.

[11] A three-domain approach originally developed by Weisser in [160], where the individual timescales τ_{1-3} represent the low-, medium- and high-temperature regimes of ignition, Section 4.4.3.

[12] A coefficient table for two temperature ranges (800 K < T < 1000 K and 1000 K ≤ T ≤ 1200 K), equivalence ratios between 0.5 and 1.5 and fuel ethanol content of up to 85 volume percent.

Fit model [69], as in this case the ignition delay times have been estimated with a reduced reaction kinetics mechanism in consideration of EGR and AFR and subsequently modeled with a modified approach. These models have been successfully utilized by several researchers without any changes [84] [86] [113], and in some studies, the ignition delay model coefficients have been modified in order to yield more accurate knock prediction results, as summarized in [34]. Additionally, Syed and Hoepke recently proposed correlations including the effects of a dilution factor x_d and the EGR rate respectively. However, they did not take the possible presence of excess air into account. Vancoillie included both a dilution factor f and the equivalence ratio in his correlation. However, no details on the calibration methodology were given in the paper. Chen proposed a similar, but simpler approach. Beccari developed a correlation for the knock onset prediction of propane, gasoline and their mixtures.

Some modifications of the base model equation were proposed by Schmid, who empirically considered turbulence effects occurring with changing engine speed as well as the presence of hot-spots. Lafossas [94] used the Douaud and Eyzat correlation for the ignition delay time in his knock model. However, he stated that the thermodynamic conditions in the compression and combustion strokes evolve strongly and quickly and therefore the original model cannot predict the knock delay directly. Instead, the estimated ignition delay time creates a 'precursor' species that is transported with the flow in the combustion chamber. When a limiting value of this precursor is reached, auto-ignition occurs. Furthermore, corrections to the model were made to take effects of the AFR and mixture dilution into account. As the model proposed by Lafossas was originally developed for the use in the course of 3D CFD simulations, enabling the prediction of the location of knock occurrence, Bougrine [18] reduced it for the use in 0D/1D engine simulations by integrating over the entire unburnt mixture instead of locally. In a later publication [127], Richard extended the approach for the use with highly downsized engines running on fuels with high ethanol content. Finally, it should be remarked that some of the models discussed in this section, e.g. the models proposed by Franzke and Worret, utilize additional empirical correlations based purely on measurement data in order to account for effects observed in the course of experimental investigations. In this context, recent approaches [56] further propose the use of 3D CFD RANS simulations for the calibration of a simple mathematical mixture inhomogeneities model that, in combination with the Livengood-Wu integral in Equation 2.11, is used for the 0D/1D simulation of the knock limited spark advance.

2.3.2.3 Conclusions

Overall, this review of the models for knock simulation shows that a lot of effort has been put into the development of a reliable 0D/1D knock model. However, the huge number of researches that have worked and are still working on this topic as well as the often reported need to modify the model coefficients for the simulation of different engine configurations reveal that **the regularly communicated poor knock prediction accuracy, is caused by a problem that still has to be identified and solved**, which is also demonstrated in Figure 2.6. The evaluation reveals that the pre-reaction state of the mixture estimated with the popular model proposed by Worret [168] for measured single cycles (see Chapter 3) changes significantly with the engine speed instead of being constant, thus indicating a problem regarding the main assumptions of the Livengood-Wu approach presented in Section 2.3.2.1.

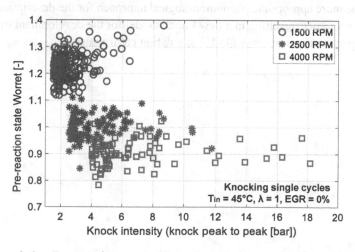

Figure 2.6: Pre-reaction state of the unburnt mixture at measured knock onset estimated with the model proposed by Worret [168] at different engine speeds.

More importantly, the known issues with the model accuracy, combined with the fact that the majority of the 0D/1D knock models are very similar, Table 2.1, suggests that major changes in the simplified approaches for modeling the real chemical and physical processes are needed in order to improve the knock prediction performance. This on the other hand requires the better understanding of the processes that lead to knock in real engines.

To this end, **it remains unclear to the author why the remarks on the limits of the Livengood-Wu integral and the appropriateness of the estimation of ignition delay times with simple Arrhenius-type equations discussed in Section 2.3.2.1 have been known and yet ignored by the majority of the researchers for more than 60 years.**

Currently, **the most promising approach for developing a new, reliable and fully predictive knock model seems to be to start from scratch and use all powerful tools available today,** such as measurement of single engine cycles, kinetic simulations of the unburnt mixture with detailed reaction kinetic mechanisms at in-cylinder conditions and the use of refined 0D/1D models to estimate the values of the knock model inputs. This shall enable the formulation of a new, more appropriate phenomenological approach for the description of knock occurrence, resulting in a new knock model for the development of future engine concepts within a 0D/1D simulation environment.

3 Experimental Investigations and Thermodynamic Analysis

For the development of a new 0D/1D knock model, the occurrence of knock has been extensively investigated on an engine test bench. To this end, a homogeneously operated direct injection spark ignition single-cylinder research engine featuring external boosting, low-pressure exhaust gas recirculation and a tumble generation device was used. This chapter briefly presents the experimental setup and describes the measurement data processing. Additionally, the effects of factors known from previous studies to influence knock are analyzed. Finally, the commonly used knock modeling approach presented in Section 2.3.2 is applied to measured single cycles, aiming at identifying reasons for its well-known poor prediction performance.

3.1 Experimental Setup Overview

The engine used in the course of the experimental investigations features external boosting, low-pressure exhaust gas recirculation, and a tumble generation device, Table 3.1 [28]. The layout of the cylinder head and combustion chamber dome is shown in Figure 3.1. The engine features a central injector, which is located between the intake valves. A spark plug with a heat value of eight (as proposed by the manufacturer NGK) is mounted between the exhaust valves. In order to guarantee realistic boundary conditions during the measurement campaign, the engine is equipped with an exhaust throttle to simulate the backpressure of the turbine. In the case of boosted operation, the average exhaust backpressure is controlled so that it equals the intake pressure [51]. The tumble generation device has been used at 50 % actuation during the majority of the performed investigations. Detailed information about the resulting level of charge motion and the flow performance can be found in [82].

The extraction point of the low-pressure exhaust gas recirculation system is located downstream of the throttle. The EGR inlet is positioned upstream of the external charging system (three-stage roots blower). In order to avoid condensation of water in the EGR cooler and the intake system, the cooler outlet

© Springer Fachmedien Wiesbaden GmbH, part of Springer Nature 2019
A. Fandakov, *A Phenomenological Knock Model for the Development of Future Engine Concepts*, Wissenschaftliche Reihe Fahrzeugtechnik Universität Stuttgart, https://doi.org/10.1007/978-3-658-24875-8_3

temperature has been set to 70 °C. The EGR rate is determined by a CO_2 measurement in both the exhaust and the intake systems and can be calculated with Equation 3.1, based on the measured concentrations and mass flows [28]. The composition of the exhaust gas is estimated with a flame ionization detector (HC), a paramagnetic oxygen analyzer (O_2), an infrared gas analyzer (CO and CO_2), a chemiluminescence analyzer (NO_x) and a TSI EEPS™ (PN) [51]. Additionally, the engine has been equipped with a conventional catalyst from a gasoline passenger car in order to enable the investigation of operation with catalyzed exhaust gas. Thanks to the high volume of the catalyst and thus low space velocities, a high conversion rate could be achieved. During catalyst operation, the components in the exhaust gas have been measured both upstream and downstream of the catalyst.

Table 3.1:   Engine specification overview [51].

Stroke (s)	90.5	mm	Valve train	DOHC (4V)	-
Bore (D)	75	mm	Exhaust valve opening[13]	13 (bBDC)	°CA
s / D	1.207	-	Exhaust valve closing[13]	10 (bTDC)	°CA
Engine displacement	399	cm³	Intake valve opening[13]	15 (aTDC)	°CA
Connection rod length	152	mm	Intake valve closing[13]	30 (aBDC)	°CA
Piston-pin offset (thrust side)	0.5	mm	Injection system	Bosch HDEV 5.2 DI solenoid multi-hole	-
Compression ratio	10.9	-	Fuel pressure	200	bar

For the data acquisition, the engine has been equipped with several thermocouples, pressure transducers, and exhaust gas measurement extraction points. The intake, exhaust, and cylinder pressure data are recorded by an indication system with a sampling rate of 0.1 °CA. All thermocouple temperatures, low-frequency pressure transducer data, and exhaust gas measurement concentrations are averaged over 30 seconds [28].

---

[13] Referred to 1 mm lift.

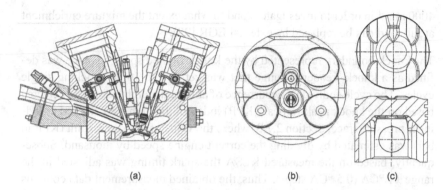

(a)   (b)   (c)

Figure 3.1:   Single-cylinder engine layout. (a) sectional view of cylinder head, (b) combustion chamber dome and (c) piston geometry for the compression ratio of 10.9 [51].

Prior to the experimental investigations, the valve timing of the engine has been optimized in order to minimize the internal EGR fraction as well as the gas exchange losses and subsequently kept constant. All test bench measurements have been performed with the same gasoline fuel with a research octane number of 96.5 and an ethanol content of 9.6 % (volume) [51]. The injection timing and pattern have not been changed during the measurement campaign.

$$x_{EGR} = \cfrac{1}{1 + \cfrac{Y_{CO_2,Exh} - Y_{CO_2,Int}}{Y_{CO_2,Int} - Y_{CO_2,Air}} \cdot \cfrac{\dot{m}_{Exh}}{\dot{m}_{Air}} \cdot \cfrac{M_{Air}}{M_{Exh}}} \qquad \text{Eq. 3.1}$$

x_{EGR}	EGR mass fraction [-]
$\dot{m}_X$	mass flow of the component X [kg/h]
M_X	mole mass of the component X [kg/kmol]
Y_X	volume fraction of the component X [-]

In the course of the measurement campaign, spark timing sweeps in the region of the knock limited spark advance have been carried out at different engine speeds, EGR rates (incremented by 5 %) and air-fuel equivalence ratios at a constant indicated mean pressure of 16 bar. Variations of the inlet temperature (incremented by 10 °C) as well as the charge motion level (open and closed tumble actuator) have been performed too. Additionally, measurements at a constant exhaust gas temperature and an AFR of 0.85 have been conducted at

4000 min^{-1} in order to investigate if and to what extent the mixture enrichment at full load can be replaced by external EGR [28].

The knock boundary represented by the knock limited spark advance was defined as a knock frequency range that was set to 4 to 10 % knocking single cycles. The cycle-individual occurrence of abnormal combustion was detected based on the knock peak-to-peak (KPP) index of the high-pass filtered in-cylinder pressure trace, Section 2.1.3, where the value of the knock limit (KPP in bar) was estimated by dividing the current engine speed by thousand. Subsequently, based on the measured KLSA the spark timing was adjusted in the range ±1 °CA (0.5 °CA steps). Thus, the obtained measurement data contains operating points with light to severe knock occurrence [51].

For the generation of additional data sets for the validation of the knock model, the engine has been modified by changing the cylinder head with one featuring tumble runners. Furthermore, the compression ratio has been increased to 11.8 by installing a new piston and the single-cylinder has been operated without sheets in the intake port and at an open tumble actuator position [28]. Thus, this second configuration differs significantly from the engine depicted in Figure 3.1 [51]. Again, variations of the EGR rate at different engine speeds have been performed; however, the engine load has been varied too, covering the range between 12 and 20 bar indicated mean pressure.

Generally, the performed investigations have revealed that despite the slower burning velocity, an earlier center of combustion and hence higher full load efficiency can be achieved with EGR (engine load- and speed-dependent). Further information about the experimental setup, an overview of the test matrix as well as a detailed measurement data analysis is available in [54] and [28]. The following sections deal with the preprocessing and the evaluation of measurement data in the context of the development of a new 0D/1D knock model.

3.2   Measurement Data Processing

A thermodynamic analysis of the combustion process has to be carried out prior to the evaluation of the available measurement data. In this work, over 200 000 single cycles were evaluated in total. Because of the huge amount of measurement data, the preprocessing as well as the thermodynamic analysis itself and the evaluation of the results have been completely automated in MATLAB [110]. Furthermore, the code was optimized to run on four CPU cores, resulting in the simultaneous calculation of four engine cycles. The following sections briefly discuss all prerequisites for a high-quality pressure trace and combustion process analyses.

3.2.1 Knock Onset Detection

In course of the performed test bench measurements at the knock boundary, the knock limited spark advance was defined at a knock frequency range between 4 and 10 % based on the knock peak-to-peak as a characteristic number of the knocking combustion, Sections 2.1.3 and 3.1. Thus, the available measurement data contains information which single cycles include a knock event. However, the evaluation of the knock integral performance and the identification of approach weaknesses require the additional detection of the knock onset of the knocking single cycles. Numerous investigations have been carried out on this topic, resulting in various knock onset detection methods based on the measured cylinder pressure trace, its first and third derivatives as well as the net heat release [1] [3] [6] [13] [16] [20] [21] [22] [23] [31] [32] [33] [36] [55] [70] [87] [88] [99] [100] [152] [168]. An overview including the most important characteristics of the different methods can be found in [23] and [168]. In this work, the knock onset was estimated as the location of the minimum of the third derivative of the filtered cylinder pressure, as proposed in [31] [32] [33]. This method is simple to implement, reliable and in the course of evaluations of the available methods, it has proven to yield accurate results for the available measurement data.

3.2.2 Pressure Trace Analysis

The combustion process was analyzed by performing a pressure trace analysis, Section 2.3.1, with the FVV cylinder module [66] as well as the FKFS UserCylinder [125]. Generally, the PTA estimates the heat release rate from the measured pressure trace based on the basic concepts of thermodynamics discussed in Chapter 2.3.1.1. Additionally, the two zones' temperatures (burnt and unburnt, Section 2.3.1.2) for both the mean cycle (averaged over 300 consecutive single cycles) and each of the single cycles have to be calculated.

Prior to the thermodynamic analysis, all measured inlet port, outlet port and cylinder pressure profiles have to be filtered. This is particularly important for the measured single cycles, as their cylinder pressure traces contain partially extreme fluctuations resulting from the occurrence of knock. On the other hand, it has to be ensured that as little information about the knock event as possible is filtered out. In this work, the measured pressure profiles were filtered with a second-order Butterworth filter, as suggested in [85] [126] [161]. The PTA enables the assessment of parameters such as combustion start and end, burn duration and center of combustion. Furthermore, a high-quality PTA is essential for the calibration of the two-zone combustion model, Section 2.3.1.2. The accuracy of the analysis is also of crucial importance for the further investigation of knock, because of the phenomenon's sensitivity to changes in the temperature of the unburned gases.

3.2.3 PTA Model Calibration

Whereas the geometry of the engine is known and thus can be used directly for the parametrization of models requiring values of geometrical parameters (e.g. piston motion model), the cylinder wall temperatures (cylinder head, piston, liner and inlet / exhaust valves) as well as the heat transfer coefficients have to be estimated prior to the combustion analysis. For evaluating the wall heat transfer coefficient from the combustion chamber gases to the cylinder walls, both the Bargende [8] and Woschni [169] wall heat transfer approaches were used in the course of this work. Because of the diverse variations of the operating conditions investigated on the engine test bench, a combustion analysis performed with imposed constant wall temperatures is expected to result in significant PTA calculation errors. This is because the wall temperatures change considerably with operating parameters such as engine speed and load as well as the EGR rate [8] [10] [169]. In order to account for these effects, the heat transfer on the outside of the cylinder components was modeled as well. This involved the estimation of the heat transfer coefficients at the interfaces head / coolant, liner / coolant, piston / oil and liner / oil. The oil and coolant temperatures are available from the measurement data.

Additionally, blowby losses were accounted for by evaluating the mass flow rate through an orifice resulting from a pressure gradient, as proposed in [65]. Furthermore, in reality the exhaust gas of an internal combustion engine contains complete and incomplete combustion products. The measured concentrations of incomplete products enabled the calculation of the fraction of the fuel's chemical energy not fully released inside the engine during the combustion process, also referred to as combustion efficiency [9] [65] [75], and its consideration in the course of the performed combustion analysis, see Section 3.2.6.

3.2.4 Pressure Pegging

The cylinder pressure signal is measured with piezoelectric pressure transducers, which can only record relative values. Thus, the measured values have to be related to the absolute pressure. This process is commonly referred to as pressure pegging. Müller has listed a series of pressure pegging methods in [116]. Furthermore, it has to be accounted for peg drift, which is defined by the changes in the sensor's pressure offset from one cycle to the next determined by the pegging method [97]. Since the available measured pressure data are shifted by the magnitude of the peg drift, all operating condition variations contained in the pegged data are influenced by this quantity, which ideally should be zero. This is particularly important for the analysis of single cycles. For the evaluation of the measurement data in this work, it was assumed that the pressure peg values of the single cycles equal the peg of their corresponding averaged cycle. Hence, pressure pegging has only been performed for the measured averaged cycles. Thus, the potential occurrence of peg drift is ignored, as the drift is not expected to rise significantly during the measurement of the 300 consecutive single cycles that compose each averaged cycle. Furthermore, the occurrence of knock and the resulting pressure fluctuations can cause a thermal shock of the piezoelectric transducers and thus temporarily lead to errors in the measured cylinder pressure values [75] [88] [93] [98], which would be reflected in the estimated pressure peg. The recovery periods of the transducers typically contain both a linear and exponential components and continue throughout the engine cycle, except at very low engine speeds. Nevertheless, the assumption that the average-cycle pressure peg is also valid for the corresponding single cycles has to be kept in mind during the measurement data evaluation.

Regarding the pressure pegging, it has to be remarked that at the PTA calculation start (typically shortly after IVC [65]), a shift of just 50 mbar at an absolute pressure of 1 bar results in a change in the starting temperature (in K) of 5 %[14], see Section 3.2.6. Considering the exponential influence of temperature on the mixture ignition delay, Section 2.3.2.1, it becomes clear that the investigation of knock requires very high quality of the measurement data and very accurate pressure pegging. For this reason, the pressure pegging was performed based on both the measured crank angle resolved inlet port pressure and the cumulative net heat release criterion [65], which assumes that no heat is released during the compression phase before the start of combustion. Consequently, in this phase the cumulative net heat release has to equal zero. To this end, the pressure level is iteratively adjusted, yielding the difference between the absolute and the measured relative pressures. The ensuing comparison of the values obtained with the two pressure pegging methods revealed no considerable differences, thus ensuring the high quality of the PTA results.

3.2.5   Cylinder Charge Estimation

Performing a PTA requires the knowledge of the cylinder mass at the beginning of the high-pressure phase (at inlet valve close). The cylinder charge is typically composed of air, fuel and exhaust gas (internal EGR, if relevant also external high- or low-pressure EGR), although prior to the measurement campaign on the engine test bench, experimental investigations were performed for setting up the valve timing so that as little internal EGR as possible is present in the cylinder. The fuel and air fractions can be estimated based on the measured AFR, fuel and air mass flows (two out of the three parameters is needed, Section 3.2.6). The internal exhaust gas mass fraction can be estimated with a well-calibrated 1D engine flow model. Alternatively, a charge exchange analysis can be performed (also known as three pressure analysis, TPA), which requires that crank angle resolved pressure data on both inlet and outlet sides as well as the valve lift curves and flow coefficients are available. In this case, the exhaust gas mass is obtained by evaluating the mass conservation equation for the open thermodynamic system "combustion chamber", as discussed in Section 2.3.1.1. In this work, the internal EGR was estimated in the course of charge exchange analyses.

---

[14] This is true only if the temperature dependence of all gas properties is negligible, which is a valid assumption for the temperature range typical at IVC [161].

3.2.6   Iterative Adaption of the Cylinder Mass

The in-cylinder boundary conditions at calculation start are decisive for the estimation of the heat release rate from the measured pressure trace. These initial conditions can be calculated with the ideal gas law, Equation 2.3, if the cylinder mass and its composition as well as the exact value of the in-cylinder pressure are known at calculation start, Sections 3.2.4 and 3.2.5. As already mentioned, the mass of fuel and air can be estimated based on the measured AFR, fuel, and air mass flows. Hence, for the calculation of the in-cylinder initial conditions, generally more measured values are available than actually needed. This fact poses the question, which measurement variables are the most trustworthy.

Due to the high number of potential sources of error in the fuel and particularly air mass flow measurements, the in-cylinder pressure signal, together with the models needed for performing a combustion analysis (e.g. wall heat transfer), are commonly assumed to be the most accurate parameters available for the estimation of the in-cylinder initial conditions [65] [161]. Consequently, the cylinder mass at calculation start is adjusted iteratively, so that it accurately corresponds to the measured pressure profile. This calculation, known as 100 %-iteration, is performed by evaluating the energy balance of the system as shown in Equation 3.2, where the converted fuel energy estimated from the measured pressure signal under consideration of the combustion efficiency, Section 3.2.3, must correspond to the energy supplied to the cylinder [9] [65] [68] [161]. In this context, it has to be remarked that any inaccuracies in the measured concentrations of incomplete products needed for the combustion efficiency estimation will result in an error in the adapted cylinder mass.

$$EB = 100\% \cdot \left(\int_{Combustion\ start}^{Combustion\ end} \frac{dQ_B}{d\varphi} d\varphi \right) / \left(m_f \cdot H_u \cdot \eta_c \right) \qquad \text{Eq. 3.2}$$

EB energy balance

m_f cylinder fuel mass [kg]

η_c combustion efficiency [-]

Different 100 %-iteration types exist, depending on the working principle of the engine. Generally, it is assumed that either the measured fuel mass, air mass or the AFR is completely trustworthy and is thus kept constant during the cylinder mass adoption. Consequently, the respective variable(s) left is / are adjusted iteratively so that the energy balance equation is satisfied. As it is commonly accepted that the measurement of the AFR performed by Lambda sensors in the typical SI engine operation range is very accurate, the AFR is assumed constant for the 100 %-iteration of the cylinder mass in SI engines. While this is generally true, it has to be considered that in case of some SI engine operation strategies, for example scavenging in combination with mixture enrichment at low engine speeds (aiming at post-oxidation in the exhaust manifold), the fuel mass should be assumed constant for the 100 %-iteration [161]. However, in this work such strategies are not of relevance.

As is the case with the pressure pegging discussed in Section 3.2.4, in this work it is assumed that the adapted cylinder mass values of the single cycles equal the adapted mass of their corresponding averaged cycle. The reason for this is that the pressure fluctuations resulting from knock as well as the possible occurrence of thermal shocks could lead to partially significant errors in the estimated cylinder mass of some single cycles. However, the assumed neglect of the cycle-resolved cylinder mass fluctuations has to be kept in mind during the measurement data evaluation.

3.3   Analysis of Factors Influencing Knock

Figure 3.2 summarizes all factors that are commonly assumed to be of importance for the accurate prediction of knock occurrence in the context of the knock integral equation. These can be divided into two basic categories: chemical and thermodynamic effects. The chemical effects, as discussed in Section 2.3.2, refer to the influences of different parameters on the ignition delay of air-fuel mixtures (e.g. exhaust gas, AFR, and fuel composition), the auto-ignition behavior, as well as the relevance of the negative temperature coefficient (NTC) zone. As shown in Figure 3.2, these effects are reflected in the estimation of the ignition delay times, which will be discussed in detail in Chapter 4. The following sections aim at the investigation of the thermodynamic effects on knock based completely on measured single cycles, so that the information present in the individual engine cycles that is not contained in the corresponding representative averaged cycle is considered.

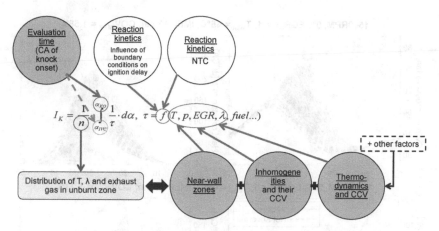

Figure 3.2: Overview of factors influencing the prediction of knock with the knock integral given by Equation 2.17.

3.3.1 Thermodynamic Effects

The chemical processes and thus the knock behavior are influenced significantly by the thermodynamic conditions in the unburnt mixture, as depicted in Figure 3.2. Typically, the values of the boundary conditions in the unburnt mixture (temperature, pressure, exhaust gas fraction, mixture composition) are used as knock model inputs, as discussed in Section 2.3.2. Thus, they influence directly the estimated ignition delay as well as the progress of the Livengood-Wu integral's value. The unburnt temperature as well as the cylinder pressure is in turn strongly affected by the position of the combustion start and the burn duration. Furthermore, the occurrence of cycle-to-cycle variations, which is typical for SI engines, results in significant fluctuations of the knock model inputs' values as well as the stated parameters that influence them, as shown in Figure 3.3. Obviously, the knocking single cycles in Figure 3.3 are the ones that burn faster-than-average. Furthermore, the pressure and thus the unburnt temperature levels during combustion of these cycles are generally higher. However, a clear distinction of the critical cycles is not possible, as the regions of the knocking and non-knocking cycles overlap. Hence, a criterion for identifying the knocking cycles cannot be derived based on these observations.

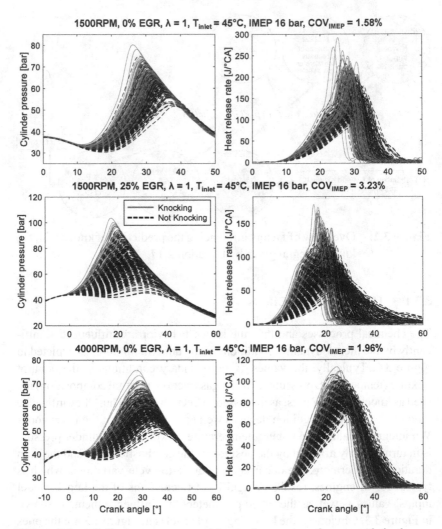

Figure 3.3: Cylinder pressures and heat release rates of measured single cycles.

Figure 3.4 shows the fluctuations of the indicated mean pressure at various operating conditions. Obviously, these increase with the EGR rate, despite the advanced spark timing and thus earlier center of combustion in this case. Hence, the influence of mixture dilution on the burn duration resulting in combustion that is more unstable predominates the stabilizing effect of earlier

MFB50-points. Furthermore, increasing the engine speed counteracts the effects of EGR on the CCV. Similarly, the influence of the AFR on the CCV predominates the EGR effects, as it affects the combustion stability dramatically. The combustion of rich mixtures is very quick and stable, whereas excess air results in very unstable combustion, similarly to diluting the mixture with exhaust gas.

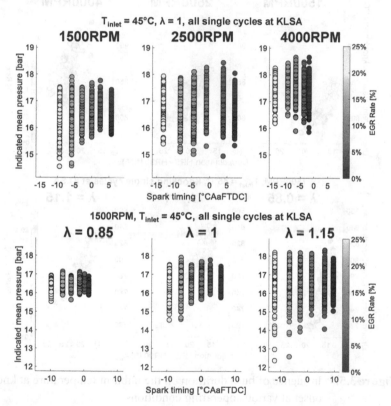

Figure 3.4: Indicated mean pressure fluctuations at various operating conditions.

Overall, the effects causing a rise in the CCV level lead to more pronounced fluctuations of the knock model inputs. This can potentially cause problems for the 0D/1D knock prediction, as the higher fluctuation level is not reflected in the corresponding averaged working cycles. Hence, the simulation of single cycles with the model presented in Section 2.3.1.3 might be inevitable for achieving an accurate prediction of the knock boundary in this case, resulting in a significant computational time increase. These observations require additional investigations on this topic with the new knock model prior to its validation.

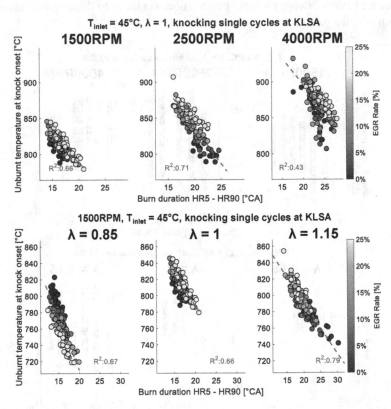

Figure 3.5:	Influence of burn duration on the unburnt temperature at knock onset at various operating conditions.

The unburnt mixture temperature at knock onset correlates with the burn duration, as shown in Figure 3.5. Generally, there is no significant difference in the burn duration of single cycles at 0 % and 25 % EGR. Hence, the laminar flame speed decrease caused by the exhaust gas, Equation 2.9, largely compensates the earlier and thus a priori faster combustion, where this trade-off is also influenced by other parameters, such as the engine speed. However, it should be remarked that the investigated burn duration (HR5 – HR90) excludes the ignition delay as well as the late combustion phase, which are both expected to be affected considerably by the presence of exhaust gas. Increasing the engine speeds results in a rise of the burn duration in degrees crank angle in general as well as in a perceptibly longer combustion with EGR. The effect of the AFR on the combustion duration is very pronounced too and is related to the slower flame speed in case of lean mixtures.

It is of common knowledge that the unburnt temperature level rises with the engine speed due to the shorter working cycle duration in the time domain and the higher cylinder wall temperatures [75]. Because of the correlation between the burn duration and the unburnt mixture temperatures at knock onset shown in Figure 3.5, the latter are also primarily influenced by engine speed and the AFR. All changes in the operating conditions that result in longer combustion durations cause the unburnt temperature at knock onset to decrease, as in this case, the peak unburnt temperatures decline as well.

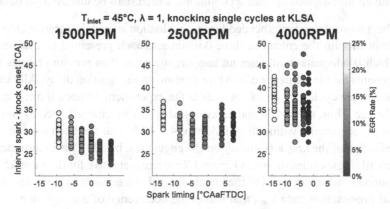

Figure 3.6: Effects of EGR and engine speed on the interval between spark and knock onset.

Additionally, the evaluation in Figure 3.6 shows that the duration of the interval between spark timing and knock onset in degrees crank angle increases with engine speed, which also results from the already discussed operating parameter's effect on the burn duration. However, the presence of exhaust gas also extends this interval, despite the observed absence of considerable EGR influence on the burn duration HR5 – HR90 in Figure 3.5. This reveals that

EGR results in longer ignition delays (interval spark – HR2 and spark – HR5) as well as presumably in a change in the MFB-points at knock onset, which will be discussed in detail in Section 3.3.3.

3.3.2   General Knock Integral Prediction Performance

One of the main goals of this work is to understand the source(s) and behavior of the knock integral error discussed in Section 2.3.2 as well as to identify possible causes for the often-reported bad knock prediction performance of the phenomenological approach. As a beginning, the pre-reaction states I_k of the unburnt mixture of measured knocking single cycles estimated with Equation 2.17 can be evaluated at a specified MFB-point representing the constant latest possible point where knock can occur assumed by the knock integral as discussed in Section 2.3.2. The corresponding results are shown in Figure 3.7 for MFB85. Obviously, the calculated values of I_k at MFB85 change significantly with all investigated operating conditions, which shall be discussed in detail.

The ignition delay times needed for this evaluation at each integration step are obtained with the enhanced three-domain approach presented in Chapter 4, which is adequate in all relevant temperature as well as pressure ranges and accounts for the influence of mixture dilution on the ignition delay. All evaluated single cycles are knocking ones at the experimental knock limited spark advance. Thus, the corresponding averaged cycles are characterized by similar knock frequencies within the window of 4 – 10 % that has been used for the definition of the knock boundary on the engine test bench. This in turn means that all single cycles shown in Figure 3.7 are representative for the same knock boundary, which in terms of the knock integral corresponds to a constant critical pre-reaction state $I_{k,crit}$ that marks the occurrence of auto-ignition in the unburnt mixture. Assuming that the critical state, as marked in Figure 3.7, equals one and accounting for the CCV (e.g. by setting the knock boundary in the range between 0.95 and 1.05), it is obvious that at the lowest investigated engine speed, none of the cycles at 25 % EGR is predicted to be a knocking one, as even at the assumed latest possible point where knock can occur of 85 % mass fraction burnt, $I_{k,crit}$ is not reached. Similarly, at 4000 min^{-1}, except for a couple of cycles without EGR, all investigated points correspond to no knock occurrence. Likewise, for the case of rich combustion, only a handful of the single cycles is predicted to knock. These results disagree with the fact that all these cycles are knocking ones and represent the same experimental knock boundary.

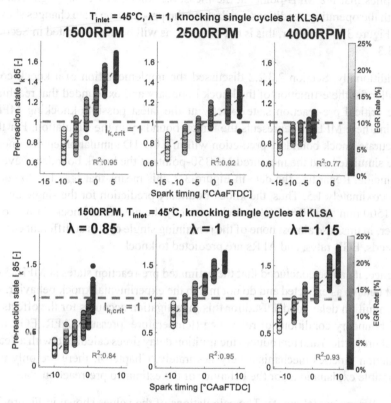

Figure 3.7: Pre-reaction state of the unburnt mixture at MFB85 at various operating conditions estimated with the knock integral.

Alternatively, it can be assumed that $I_{k,crit}$ at the experimental knock boundary equals the smallest cycle-individual value in Figure 3.7 of approximately 0.2. In this way, all measured single cycles are predicted to knock, which is in agreement with the experimental results. However, as discussed in Section 2.3.2.2, the pre-reaction state curve is always rising, because the ignition delay times and thus the integrand values in the knock integral equation are always positive. Hence, in case of a $I_{k,crit}$ of 0.2, the single cycles with high pre-reaction states at MFB85, e.g. at 1500 min^{-1} and 0 % EGR, will reach the assumed critical I_k value at very early MFB-points, some of them possibly even before the combustion start, which would mean pre-ignition. Clearly, this observed behavior of the estimated I_k values at MFB85 is not credible. Furthermore, it implies that the MFB-points at the measured knock onset vary significantly with the operating conditions similarly to the pre-reaction state changes shown in Figure 3.7. However, this is not the case, as will be demonstrated in Section 3.3.3.

Additionally, Section 2.3.2.2 discussed the implementation of a knock controller for the estimation of the knock boundary and expounded that reaching the critical pre-reaction state exactly at the latest possible knocking MFB-point, here MFB85, represents the knock boundary in the simulation. For the accurate knock boundary prediction within a 0D/1D simulation environment, the simulated and the measured MFB50-points at the knock boundary have to coincide. Regarding the data in Figure 3.7, this means that $I_{k,crit}$ has to equal approximately 1.5. Thus, the knock boundary prediction for the single cycles at 1500 min^{-1} and 0 % EGR matches exactly the measured knock limit. However, in this case almost none of the remaining single cycles at different engine speeds, EGR rates, and AFRs are predicted to knock.

Hence, it can be concluded that the estimated pre-reaction states at MFB85 do not behave as expected and do not match the experimental knock behavior. As the ignition delay model used for this investigation accounts for the effects of all boundary conditions of relevance (temperature, pressure, AFR, EGR and fuel properties) and reproduces the ignition delay times calculated with a detail reaction kinetics mechanism very accurately, Chapter 4, there are only two possible explanations for the behavior of the estimated pre-reaction states:

■ Wrong model inputs: The calculations of the values shown in Figure 3.7 were performed with the mean values of the unburnt mixture parameters. Hence, in the course of this investigation the influences of near-wall zones, mixture and temperature inhomogeneities have been neglected, which could explain the behavior of I_k. The exact spot in the unburnt zone, where auto-ignition occurs and causes knock, is expected to have local temperature, AFR and EGR fraction different from the mean unburnt values as discussed in detail in Chapter 3.3.6.

■ Incorrect knock modeling approach: Flaw(s) of the knock integral itself, Equation 2.17, and / or the assumptions that it is based on can result in the unexpected behavior shown in Figure 3.7.

Hence, the main task at this point is to find out which of these two options (or maybe both) cause(s) the observed behavior of Ik and thus the poor prediction performance that 0D/1D knock models are known for.

3.3.3 Knock Occurrence Criterion

It is commonly assumed that knock occurs because of local auto-ignition(s) in the unburnt mixture, Sections 2.1.2 and 2.3.2. However, the prediction of local auto-ignition where $I_k = I_{k,crit}$ is not sufficient for the reliable calculation of the knock boundary, as the occurrence of this phenomenon does not necessarily result in knock [88] [89]. Hence, an additional criterion for occurrence of knock resulting from the predicted auto-ignition is needed. To this end, the evaluation of the Livengood-Wu integral is performed to a constant MFB-point, as it is assumed that an auto-ignition after this point does not result in knock because of the small unburnt mass and volume fractions left, Section 2.3.2.2. This end-of-integration point, also referred to as evaluation time of the knock integral, is always constant and typically in the range MFB75 to MFB85, with some knock models assuming that an auto-ignition after MFB95 can also result in knock, as discussed in Section 2.3.2.2.

Figure 3.8 shows the estimated knock onset of measured knocking single cycles at various operating conditions. Obviously, the MFB-points at the experimental knock onset do not only vary because of the cycle-to-cycle variations, but also change systematically with the operating conditions. The knock onset at high EGR rates always slightly shifts to earlier MFB-points, meaning that more mixture auto-ignites when EGR is employed. Additionally, increasing engine speeds result in later knock onsets. Rich mixtures knock later too. Similarly to the effect of EGR, excess air causes a shift of the knock onset to earlier MFB-points, suggesting that there is a general effect of mixture dilution on the knock onset, which is probably also related to the typically higher cylinder mass as well as longer burn duration in this case. Interestingly, although high EGR rates result in unstable combustion and thus high levels of the CCV as shown in Figure 3.4, the presence of exhaust gas does lead to less fluctuations of the MFB at knock onset.

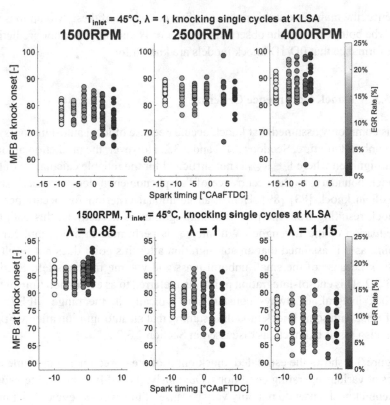

Figure 3.8: Mass fraction burnt at measured knock onset of knocking single cycles at various operating conditions.

Overall, the investigation performed in this section shows that the assumed constant end-of-integration point is not feasible and clearly causes prediction errors. Thus, this is one of the reasons for the poor prediction performance that 0D/1D knock models are known for. Because of the significant changes in the MFB-points at knock onset with the operating conditions, only a cycle-individual knock occurrence criterion that estimates the latest MFB-point where knock can occur as a function of the operating conditions can improve the accuracy of the knock prediction.

3.3.4 Top Land Influence

Previous studies have suggested that the mass in the piston's top land that is colder than the rest of the unburnt mass in the cylinder can flow out and cause a slowdown or even a freeze of the chemical reactions in the unburnt mixture and thus suppress knock [10] [135]. Regardless of the result of the interaction of the two phenomena (knock occurrence promotion or mitigation), it is first important to determine whether this interaction can take place. It is well known that the top land mass cannot start flowing out before the maximum cylinder pressure has been reached, as this is the earliest possible point for the mass flow direction change [10]. Thus, the question that has to be answered is how the position of maximum cylinder pressure relates to where light to moderate knock, which is commonly used to determine the knock boundary, occurs. In a SI engine running on gasoline, light to moderate knock typically occurs after MFB65 [52], see also Figure 3.8.

The MFB at maximum pressure and its relation to the measured MFB at knock onset of single cycles at various operating conditions are shown in Figure 3.9. The measured single cycles suggest that peak pressure mostly occurs later than knock onset (cycles on the left of the bisector in Figure 3.9), meaning that the maximum pressure results from knock. Increasing the engine speed, however, leads to earlier MFB-points at maximum pressure, thus causing an inversion of the relation between the points of maximum pressure and knock onset. Hence, the top land mass could have an influence on knock at higher engine speeds. In this context, the top land mass is suggested to have contributed to the observed shift of the MFB-points at knock onset towards later values at higher engine speeds, Figure 3.8. However, as these observations are engine-specific and obviously change with the operating conditions, a further investigation of the discussed interactions for the general case is reasonable. To this end, several Wiebe heat release rate functions, as presented in Section 2.3.1, were calibrated to create synthetic data and perform variations of burn duration, combustion start and thus to alter the position and MFB at maximum cylinder pressure. The variations were performed at different operating conditions, e.g. EGR rates and engine speeds. As the results are based on synthetically created data, the findings are supposed to be valid in general.

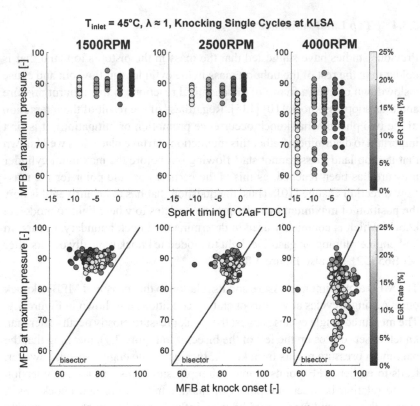

Figure 3.9: MFB at maximum pressure in relation to the MFB at measured knock onset of single cycles at various operating conditions.

Early MFB50-points typically result in a shift of the crank angle where the maximum pressure occurs. Accordingly, the maximum pressure values rise as these appear ever nearer the firing top dead center. For the simulations performed with Wiebe heat release rate functions, these interrelations are almost linear, as shown in Figure 3.10.

The amount of heat released at maximum pressure is primarily influenced by the burn duration or the duration between combustion start and MFB50 respectively, Figure 3.11 (the circle size accounts for the EGR rate, with the small circles representing 0 % and the big ones 25 % EGR). The shorter the burn duration, the more heat has been released before maximum pressure is reached, meaning that the top land mass becomes irrelevant for the occurrence of knock. On the other hand, increasing the duration between combustion start and MFB50 leads to less heat released at maximum pressure.

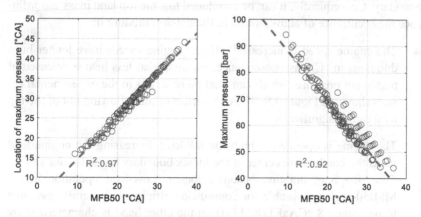

Figure 3.10: Correlation between MFB50 and the location and value of maximum cylinder pressure.

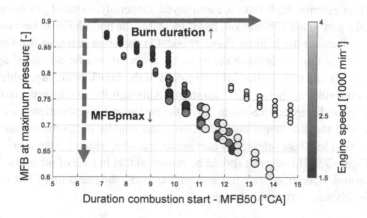

Figure 3.11: Influence of duration between combustion start and MFB50 on the heat released at maximum pressure at stoichiometric conditions.

Thus, the top land mass is generally expected to start flowing out before knock onset and can influence the occurrence of auto-ignition in the unburnt mixture if the operating point has a late combustion center, as late MFB50-points and long burn durations are concomitant. Thus, this reflection is especially relevant for engines prone to knock that operate with late spark timing at the knock boundary. Consequently, it can be concluded that the top land mass can influence the occurrence of auto-ignition in the unburnt mixture if:

- The engine speed is increased. Higher engine speeds have longer burn durations in °CA as consequence, meaning that less heat is released at maximum pressure, which can also be observed in the experimental results shown in Figure 3.9. Thus, the effect of mass flowing out of the top land can be significant.

- The engine is operated at high- or full-load. Increasing the engine load shifts the combustion center at the knock boundary towards later values. Thus, the burn duration becomes longer. Low-load operation in the MFB50-region favorable for combustion efficiency (generally assumed to be around 8 °CAaFTDC [75]) on the other hand is characterized by much faster combustion, meaning that in this case, the top land mass is not expected to be relevant for the occurrence of knock.

- High external EGR rates are employed. Generally, exhaust gas does not take part in the combustion, but it increases the burn duration because of its effect on the laminar flame speed. On the other hand, as it takes up heat and thus cools down the cylinder, it also allows for advanced spark timings and thus earlier MFB50-points at the knock boundary, which in turn results in shorter burn durations. Although the two effects are contrary as discussed in Section 3.3.1, the performed investigations have revealed that the longer burn duration effect slightly predominates and results in less heat released at maximum pressure, which is also implied in Figure 3.9. However, it should be remarked that in case of other configurations, operating strategies and engine loads, the prevailing EGR effect in this trade-off could change.

- Fuels with low octane numbers are used. These cause a retardment of the spark timing at the knock boundary leading to later MFB50-points and longer burn durations.

3.3.5 Appropriateness of the Commonly Used Auto-Ignition Prediction Approach

In Section 3.3.3, it was found out that the knock onset varies with the operating conditions and as a result, the assumed constant latest possible MFB-point where knock can occur leads to prediction errors. However, this observation is only relevant for the assessment if an occurred auto-ignition results in knock. Additionally, a simplified approach for the prediction of auto-ignition that reproduces the behavior of the chemical processes in the unburnt mixture as accurately as possible is needed for modeling knock in 0D/1D. To this end, commonly the knock integral in Equation 2.17 is used and as discussed in Section 2.3.2.1, already its developers Livengood and Wu have noted that interfering effects might arise and impair its prediction performance. Hence, the appropriateness of the commonly used auto-ignition prediction approach itself has to be evaluated by identifying potential sources of error as well as general flaws of the approach.

To this end, the prediction error resulting from the unfeasible criterion for knock occurring because of auto-ignition ascertained in Section 3.3.3 has to be eliminated. This can be achieved by estimating the I_k values at the experimental knock onset as shown in Figure 3.12, where the ignition delay times needed at each integration step are obtained with the enhanced three-domain approach presented in Chapter 4. It accounts for the effects of all boundary conditions of relevance and reproduces the ignition delay times calculated with a detail reaction kinetics mechanism very accurately. Furthermore, as discussed in Section 3.3.2, the investigated single cycles all represent the same experimental knock boundary.

By neglecting the time span between auto-ignition and the occurrence of oscillations in the measured cylinder pressure characterizing a knock event, it can be assumed that the experimental knock onset coincides with the time of auto-ignition in the unburnt mixture. This assumption implies that all pre-reaction states at knock onset in Figure 3.12 equal the critical pre-reaction state $I_{k,crit}$ that represents the knock boundary measured on the engine test bench. The theory the Livengood-Wu integral is based on postulates that the critical pre-reaction state representing auto-ignition is always constant. Thus, all calculated pre-reaction states at knock onset have to be of the same magnitude. However, Figure 3.12 clearly shows extreme changes of the values of partially more than 100 % over the investigated operating conditions. This suggests that the commonly used knock integral is not capable of accurately estimating the progress of the chemical reactions leading to auto-ignition at in-cylinder conditions, resulting in poor knock prediction performance.

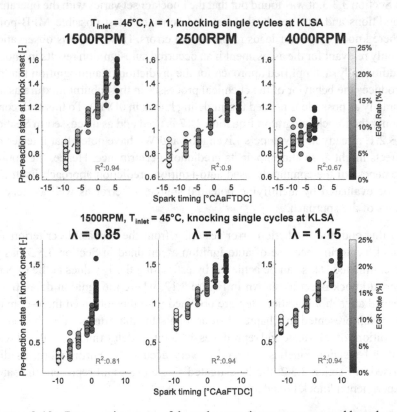

Figure 3.12: Pre-reaction state of the unburnt mixture at measured knock onset at various operating conditions estimated with the commonly used knock integral.

However, the calculations of the pre-reaction states at knock onset were performed with the mean values of the unburnt mixture parameters. Hence, in the course of this investigation the influences of near-wall zones, mixture and temperature inhomogeneities have been neglected, which could explain the behavior in Figure 3.12 and thus the poor knock integral prediction performance. In order to account for inhomogeneities effects, the local boundary conditions at the exact spot in the unburnt zone, where auto-ignition occurs and causes knock, have to be estimated and used as knock model inputs.

3.3.6 Unburnt Mixture Inhomogeneities

It is commonly assumed that cycle-individual composition (mixture) and temperature inhomogeneities in the unburnt mass result in mostly more than one exothermic centers in the unburnt mixture generally known as "hot-spots" that influence the knock behavior [88] [132] [133] [135]. Such temperature fluctuations with amplitudes of well above 10 K can exist even in a nominally homogeneous engine [132] [133]. On the other hand, the occurrence of a homogeneous auto-ignition (thermal explosion) in the end-gas is unlikely [88]. It has to be remarked that the term "hot-spots" is also used to describe locations on the combustion chamber walls such as an overheated valve or spark plug, or a glowing combustion chamber deposit [75].

In order to account for the effects of inhomogeneities, a knock model requires only the local temperature and mixture composition at the spot where knock is initiated. In the 0D/1D engine simulation, no detailed information about the exact reason for the spot formation, type, and location is available, but it is also not needed. This spot can be referred to as a "knock-spot" – a generic term for hot-spots (a glowing (part of) a component in the cylinder or a glowing fuel / oil particle), and locations in the unburnt mixture, where the boundary conditions differ from the corresponding mean values because of charge motion and turbulence influences or combustion and flame propagation effects.

Generally, 0D/1D simulations do not account for pressure fluctuations, meaning that there is no difference between the pressures in the burnt and unburnt zones in case of a two-zone combustion model, Section 2.3.1.2. Thus, a knock-spot can only have different temperature and mixture composition (AFR and exhaust mass fraction) than the mean ones for the unburnt zone. However, currently no quasi-dimensional models that can coupled with the Entrainment model presented in Section 2.3.1.2 exist for the simulation of inhomogeneities. Hence, for the evaluation of the inhomogeneities' influence on the knock integral performance, the boundary conditions at the knock-spot, which constitute the knock model inputs, have to be estimated from the available measurement data. This is a challenging task, because the values of three different local parameters have to be calculated.

To this end, it can be assumed that the local AFR as well as the local EGR fraction equal the mean unburnt zone values. Thus, only the temperature at the location in the unburnt mixture where knock is initiated has to be estimated. This assumption is feasible, because out of the three parameters, the temperature is the only one that has an exponential influence on the ignition delay times, Equation 2.12, thus being the unburnt mixture parameter that has the most pronounced effect on the knock model. Of course, the assumption results in a marginal error, which will be reflected in the calculated local knock-spot temperatures – these will differ slightly from the real values. Hence, the main task in this section simplifies to the estimation of the local temperatures at the locations in the unburnt mixture where knock is initiated for each measured knocking single cycle, which can be performed iteratively as shown in Figure 3.13.

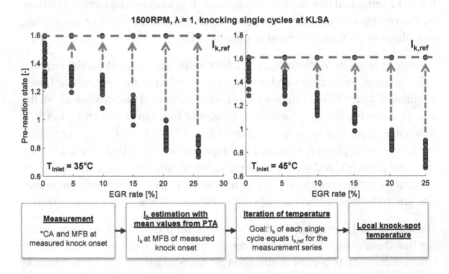

Figure 3.13: Iterative calculation of the temperature at the location in the unburnt mixture where knock is initiated.

The needed local temperature is obtained as the sum of the mean unburnt temperature and an iteratively calculated offset. The goal of the iterative calculation is to estimate the temperature offset needed so that the pre-reaction state of each single cycle at the experimental knock onset calculated with the knock integral equals a constant reference value. The reference value was selected individually for each measurement series, in this case EGR variations, as the maximum I_k achieved with the mean unburnt temperature as shown in Figure 3.13. Thus, the assumption that the critical pre-reaction state representing auto-ignition is always constant is satisfied and all calculated temperature off-set values are positive. Like in the previous sections, the ignition delay times needed at each integration step are obtained with the enhanced three-domain approach presented in Chapter 4, which accounts for the effects of all bound-ary conditions of relevance. Furthermore, as discussed in Section 3.3.2, the investigated single cycles all represent the same experimental knock boundary. As the wall heat transfer approach used in course of the pressure trace analysis significantly influences the unburnt temperature level, the calculations were performed with both the Woschni [169] and Bargende [8] wall heat transfer approaches.

The calculated temperature offsets are shown in Figure 3.14. If these are added to the unburnt temperature curves of the corresponding single cycles, all pre-reaction states at knock onset within a single EGR variation in Figure 3.13 equal the assumed constant critical pre-reaction state for the measurement se-ries. The offset behavior suggests that the temperature inhomogeneities change dramatically with the operating conditions. More importantly, many values are unrealistically high, partially well over 100 K.

Clearly, both the offset behavior (linear increase with EGR) and the absolute values are not credible, as the temperature fluctuations in the unburnt mixture are commonly reported to be between 10 K and 20 K [132] [133], which also coincides with the 3D CFD stratification analysis results reported in [28]. An-other problem is the fact the offset values in Figure 3.14 only reflect the influ-ence of EGR (individual reference value for each EGR variation). However, the pre-reaction states calculated with the mean unburnt temperature change significantly also over the AFR and engine speed as shown in Figure 3.12. Hence, the temperature offsets needed so that the calculated I_k at the measured knock onset is constant over all operating condition variations partially amount to well over 200 K (an operating point at high engine speed and EGR rate as well as fuel excess represents the extreme case). However, such unburnt tem-perature fluctuations are not possible in a real engine. This fact indicates that there is a fundamental problem with the approach used for the estimation of the temperature offsets and thus the pre-reaction states at the experimental knock onset.

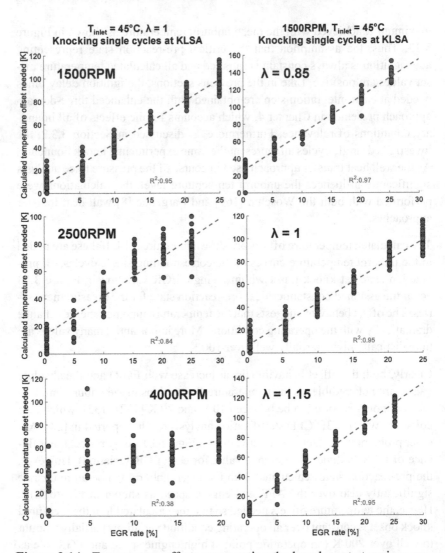

Figure 3.14: Temperature offsets representing the knock-spot at various operating conditions estimated with the knock integral.

This investigation confirms that the commonly used knock integral is not capable of accurately estimating the progress of the chemical reactions leading to auto-ignition in the unburnt mixture, as the analyses in Sections 3.3.2 and

3.3.5 have already suggested. This, together with the knock occurrence criterion based on the assumption of a constant latest possible MFB-point where knock can occur, can be stated as the reasons for the commonly reported poor knock prediction performance. As the auto-ignition prediction approach is inaccurate and does not represent the real chemical processes, at this point it remains unclear, if inhomogeneities have to be accounted for in order to achieve high knock prediction accuracy. Further investigations on this topic with an appropriate auto-ignition prediction approach are needed.

3.4 Analysis Conclusions, Problems and Limits

The analysis of the combustion process of single engine cycles generally poses challenges that are not of relevance when performing a conventional pressure trace analysis of averaged working cycles.

An inaccuracy concerning the interpretation of the measurement data could result from the lambda control, as it causes changes in the mixture composition on single-cycle basis and thus combustion fluctuations. This on the other hand results in cycle-resolved fluctuations of the fuel and air mass as well as, of course, the AFR. However, the measurement of these three parameters yields mean values over time, typically one value per average cycle, resulting in errors in the iterative adaption of the cylinder mass of single cycles. For this reason, the performed analysis of the measurement data assumes that the pressure pegs as well as the adapted cylinder mass values of the single cycles equal those of their corresponding averaged cycle. While these assumptions are feasible for reasons discussed in Sections 3.2.4 and 3.2.6, in reality peg drift, cylinder pressure fluctuations caused by knock as well as thermal shocks of the pressure transducers occur, thus causing additional marginal analysis errors resulting from the evaluated in-cylinder pressure data.

Additionally, in general the AFR is assumed constant for the 100 %-iteration of the cylinder mass of SI engines, Section 3.2.6. However, test bench setups commonly perform measurements with more than one Lambda sensors. Furthermore, it is possible to estimate an AFR value from the measured emissions concentrations with Brettschneider's formula [131]. In general, the available AFR values are never equal, and sometimes, significant deviations are present.

Hence, the choice of the most trustworthy AFR value available influences directly the iterative adaption of the cylinder mass and its results. It should be also mentioned that in case of an engine with more than one cylinders, additional uncertainties result from the cylinder-individual AFR. This, however, is not relevant for the investigations performed in this work.

In the course of the measurement campaign on the engine test bench, the knock limited spark advance was defined at a knock frequency range between 4 and 10 %. Hence, each measured operating point is characterized by a marginally different knock frequency. Strictly speaking, this means that the investigated points all represent a slightly different knock boundary. This fact has to be considered during the assessment of the new knock model's prediction performance in the course of the model validation.

A topic that has not been discussed in this chapter at all is the detection of knock onset of averaged working cycles. The fact that these contain many non-knocking single cycles poses the question if and how they have to be considered. However, as all investigations in this chapter have been performed based on single cycles for reasons discussed at the beginning of Section 3.3, the knock onset detection of averaged cycles is not of importance here. An additional uncertainty results from the fluctuations of the cylinder pressure traces caused by the occurrence of knock and their influence on the detected knock onset. Furthermore, the knock onset estimation is generally very sensitive to the type, order, and frequency of the filter used for the pre-processing of the measured in-cylinder pressure curves.

The thermodynamic analysis and the investigations performed in this chapter has provided a clear answer to the question, what causes the often-reported poor performance of all commonly used 0D/1D knock models that are based on the Livengood-Wu integral. The generally applied knock integral is not capable of accurately estimating the progress of the chemical reactions leading to auto-ignition in the unburnt mixture. Additionally, the assumed constant end-of-integration and thus latest possible MFB-point where knock can occur is not feasible and clearly causes prediction errors. **The accurate estimation of the knock boundary within a 0D/1D simulation environment requires the development of both a new auto-ignition prediction approach and a cycle-individual knock occurrence criterion.**

Additionally, it was demonstrated that long burn durations lead to less heat released at maximum pressure. In this case, the mass in the top land is expected to start flowing out before knock onset. Thus, it is supposed to have an influence on the auto-ignition process in the unburnt mixture in this case. The thermodynamic investigations of single cycles have revealed that the effects causing a rise in the CCV level lead to more pronounced fluctuations of the knock model inputs, which can potentially cause problems for the 0D/1D knock simulation. Hence, in order to achieve an accurate prediction of the knock boundary, it may be inevitable to simulate single cycles with an appropriate model. Besides, it is still unclear, if inhomogeneities have to be accounted for in order to achieve high knock simulation quality. Further investigations on these two topics are needed with newly developed, appropriate auto-ignition prediction and knock occurrence approaches.

4 Unburnt Mixture Auto-Ignition Prediction

As it is commonly assumed that knock occurs because of local auto-ignition(s) in the unburnt mixture, Sections 2.1.2 and 2.3.2 the accurate prediction of auto-ignition is the key to developing a fully predictive knock model for the 0D/1D engine simulation. This chapter aims at better understanding the chemical processes resulting in auto-ignition and developing an appropriate approach for predicting the occurrence of this phenomenon.

To this end, investigations with a detailed reaction kinetics mechanism at in-cylinder conditions are performed and the weaknesses of the commonly used Livengood-Wu integral presented in Section 2.3.2.1 are identified and examined in detail. Subsequently, a new auto-ignition prediction approach considering the effects of all currently conceivable knock suppression measures that could be employed within the framework of future SI engine concepts is developed and validated. However, first the fundamentals of reaction kinetic simulations shall be briefly discussed.

4.1 Basic Concepts of Reaction Kinetic Simulations

All reaction kinetics investigations in this work were performed with the free suite of object-oriented software tools Cantera [42] in combination with MATLAB [110]. It can be used for the kinetic simulation of combustion applications and provides an ordinary differential equations (ODE) solver for solving the stiff equations of reacting systems. This section summarizes the fundamental principles as well as the governing equations of the reaction kinetic simulations performed in this study.

The Cantera reactor represents the simplest way to model a chemically reacting system. It represents an extensive thermodynamic control volume, in which all state variables are homogeneously distributed. All system states are functions of time, resulting in an unsteady thermodynamic system. In particular, transient state changes due to chemical reactions are possible. Thermodynamic equilibrium is however assumed to govern the reactor at all instants of

© Springer Fachmedien Wiesbaden GmbH, part of Springer Nature 2019
A. Fandakov, *A Phenomenological Knock Model for the Development of Future Engine Concepts*, Wissenschaftliche Reihe Fahrzeugtechnik Universität Stuttgart, https://doi.org/10.1007/978-3-658-24875-8_4

time. Additionally, the thermodynamic system can interact with the surroundings in multiple ways and all interactions can vary as a function of time or state. Interactions of the reactor with the environment are typically defined on one or multiple walls, inlets, and outlets, similarly to the thermodynamic system "combustion chamber" presented in Section 2.3.1.1. By moving the walls of a reactor, its volume can be changed and expansion or compression work can be done by or on the system. Additionally, walls can influence the chemical reactions in the system, for example by modeling a mass transfer between the surface and the fluid. An arbitrary heat transfer rate can be defined to cross the boundaries of the reactor. Furthermore, a reactor can have multiple inlets and outlets. For the inlets, arbitrary states can be defined and fluid with the current state of the reactor exits the system through the outlets.

In addition to single reactors, interconnections of reactors into a network can be modeled too. Each reactor in a network may be connected so that the contents of one reactor flow into another, representing the exchange of matter between different thermodynamic sub-systems. Reactors may also be in contact with one another or the environment via walls, which move or conduct heat.

$$\dot{m}_{k,gen} = V \dot{\omega}_k W_k + \dot{m}_{k,wall} \qquad \text{Eq. 4.1}$$

k species index [-]

$\dot{m}_{k,gen}$ mass flow of generated species [kg/s]

$V \dot{\omega}_k W_k$ generation rate [kg/s]

$\dot{m}_{wall}$ production of homogeneous phase species on the walls [kg/s]

The state variables for a general reactor model are the mass of the reactor's contents, the mass fractions for each species, the reactor volume, and a state variable describing the energy of the system. Depending on the model, the energy of the system can be represented by the total internal energy of the reactor contents, the total enthalpy, or the temperature. The mass flow of the species that are generated through homogeneous phase reactions is given by Equation 4.1. Furthermore, a reactor always has to satisfy the principles of mass, species, and energy conservation, Equations 4.2, 4.3 and 4.4 respectively.

$$\frac{dm}{dt} = \sum_{in} \dot{m}_{in} - \sum_{out} \dot{m}_{out} + \dot{m}_{wall} \qquad \text{Eq. 4.2}$$

m	total mass of contents [kg]
$\dfrac{dm}{dt}$	total mass change [kg/s]
$\dot{m}_{in}$	mass flow through inlets [kg/s]
$\dot{m}_{out}$	mass flow through outlets [kg/s]

$$m\frac{dY_k}{dt} = \sum_{in} \dot{m}_{in}(Y_{k,in} - Y_k) - \dot{m}_{k,gen} + Y_k\dot{m}_{wall} \qquad \text{Eq. 4.3}$$

Y_k	species mass fraction [-]
$\dfrac{dY_k}{dt}$	change of species mass fraction [1/s]
$Y_{k,in}$	mass fractions of species entering the system through inlets [-]

$$\frac{dU}{dt} = -p\frac{dV}{dt} - \dot{Q} + \sum_{in} \dot{m}_{in}h_{in} - h\sum_{out} \dot{m}_{out} \qquad \text{Eq. 4.4}$$

U	total internal energy [J]
$\dfrac{dU}{dt}$	total internal energy change [J/s]
p	pressure [Pa]
V	reactor volume [m³]
$\dfrac{dV}{dt}$	volume change [m³/s]
$\dot{Q}$	total rate of heat transfer through all walls [J/s]
h_{in}	specific enthalpy of species entering the system through inlets [J/kg]
h	specific enthalpy [J/kg]

The ideal gas reactor model is of particular importance in this work, as it is used for the estimation of ignition delay times at various boundary conditions as discussed in Section 4.4.1. In this case, instead of the total internal energy U, the reactor's state variable is the temperature T [150]. For an ideal gas, the total internal energy can be rewritten in terms of mass fractions and temperature, yielding the energy conservation and thus the temperature equation for this type of models, Equation 4.5. This formulation is further simplified for a closed, adiabatic system, as in this case the inlet and outlet mass flows as well as the total rate of heat transfer through the walls are zero.

$$
mc_v \frac{dT}{dt} = -p \frac{dV}{dt} - \dot{Q} + \sum_{in} \dot{m}_{in} \left(h_{in} - \sum_{k} u_k Y_{k,in} \right)
$$
$$
- \frac{pV}{m} \sum_{out} \dot{m}_{out} - \sum_{k} \dot{m}_{k,gen} u_k
$$

Eq. 4.5

c_v	heat capacity at constant volume [J/kg/K]
$\dfrac{dT}{dt}$	system temperature change [K/s]
u_k	specific internal energy of species k [J/kg]

Reactors by themselves just define the governing equations of the investigated thermodynamic system. The implemented solver performs the integration over time.

$$
v(t) = K(p_{left} - p_{right}) + v_0(t)
$$

Eq. 4.6

v	wall velocity [m/s]
K	non-negative constant [m/s/Pa]
v_0	velocity specified as a function of time [m/s]

The reactor networks allow the interconnection of multiple sub-systems, which enables the implementation of mass flows from one reactor into another as well as heat transfer. Walls are movable objects that separate two reactors. A wall has a finite area, may conduct or radiate heat between the two reactors on either side, and may move like a piston with a velocity given by Equation 4.6, where a movement to the right is assumed to be performed with positive wall velocity.

The system volume change as a function of time due to the motion of one or more walls is given by Equation 4.7. The total rate of heat transfer through all walls can be estimated with Equation 4.8, where the heat flux through a wall connecting reactors "left" and "right" is computed with Equation 4.9. A positive heat flux implies heat transfer "left" to "right".

$$\frac{dV}{dt} = \sum_w f_w A_w v_w(t)$$

Eq. 4.7

w	wall index [-]
f_w	constant for the wall facing (± 1) [-]
A_w	wall surface area [m^2]

$$\dot{Q} = \sum_w f_w \dot{Q}_w$$

Eq. 4.8

$$\dot{Q}_w = UA(T_{left} - T_{right}) + \varepsilon\sigma(T_{left}^4 - T_{right}^4) + Aq_0(t)$$

Eq. 4.9

U	heat transfer coefficient [W/m^2/K]
ε	wall emissivity [-]
σ	Stefan-Boltzmann radiation constant [Wm^{-2}K^{-4}]
q_0	heat flux specified as a function of time [W/m^2]

$$\dot{m} = max(\dot{m}_0, 0)$$

Eq. 4.10

$\dot{m}_0$	mass flow specified as a constant or a function of time [kg/s]

A mass flow controller maintains a specified mass flow rate independent of upstream and downstream conditions, Equation 4.10. Since a reversal of the flow direction is not allowed, in case of specified negative mass flow values, the rate is set to zero. The flow is maintained even if the downstream pressure is greater than the upstream pressure, but the controller does not account for the work required to do this.

Finally, the ambient fluid surrounding the reactor network is modeled as a reservoir, which has an infinitely large volume, where all states are predefined and never change from their initial values.

4.2 Blending Rules and Reaction Mechanism Overview

Because of the complexity of detailed combustion kinetics of real fuels that are composed of a large variety of hydrocarbon components, surrogate mixtures of only few representative species are typically employed in gasoline reaction kinetics simulations, Section 2.2. The gasoline fuel used in the course of the test bench measurements (RON 96.5, E10), Section 3.1, was also selected as a base fuel for all reaction kinetic simulations performed in this work. As for ethanol-doped gasoline fuels, mixtures of n-heptane, iso-octane, toluene, and ethanol are commonly used for mimicking the target properties of the real fuel [54], the surrogate employed here is composed of these four components.

Generally, the values of the fuel RON, MON, H/C ratio and liquid density are all important for the accurate description of the real gasoline's auto-ignition behavior, Section 2.2 [28] [38] [54]. For the calculation of the H/C ratio and liquid density of surrogate mixtures, mole- and volume-fraction-based blending rules can be used, respectively [54]. The employed blending rules for the estimation of the surrogate octane numbers are founded on the investigation findings in [114] that demonstrated the need for non-linear volumetric blending of iso-octane, n-heptane, and toluene. Ethanol was blended linearly on mole fraction basis as suggested in [2]. Thus, the surrogate octane numbers can be calculated as shown in Equation 4.11 and Equation 4.12 [54]. The application of the non-linear volumetric blending rules leads to an improved prediction accuracy of both the RON and octane sensitivity (RON - MON) [54] [117].

Additionally, the estimated surrogate composition has been optimized numerically by minimizing all property differences between the real and surrogate fuels [28]. This resulted in a final surrogate composition of 45.7 % iso-octane,

13.6 % n-heptane, 30.5 % toluene and 10.3 % ethanol (mass percent) for the selected base fuel [54]. The blending rules stated in this section enable the estimation of various surrogate compositions representing different real gasoline fuels in general, so that the fuel influence on the mixture's auto-ignition behavior can be investigated in the course of reaction kinetic simulations.

$$RON = (1 - X_{Eth}) \left[100 \frac{v_{Iso}}{v_{Iso} + v_{Hep}} + 142.79 \frac{v_{Tol}}{v_{Iso} + v_{Hep} + v_{Tol}} - \right.$$

$$\left. 22.65 \left(\frac{v_{Tol}}{v_{Iso} + v_{Hep} + v_{Tol}} \right)^2 - 111.95 \frac{v_{Iso}}{v_{Iso} + v_{Hep}} \frac{v_{Tol}}{v_{Iso} + v_{Hep} + v_{Tol}} \right] + \quad \text{Eq. 4.11}$$

$$X_{Eth} RON_{Eth}$$

RON	research octane number of the surrogate [-]
RON_{Eth}	research octane number of ethanol [-]
v_C	volume fraction of component C [-]
X_{Eth}	mole fraction of ethanol [-]

$$MON = (1 - X_{Eth}) \left[100 \frac{v_{Iso}}{v_{Iso} + v_{Hep}} + 128 \frac{v_{Tol}}{v_{Iso} + v_{Hep} + v_{Tol}} - \right.$$

$$\left. 19.21 \left(\frac{v_{Tol}}{v_{Iso} + v_{Hep} + v_{Tol}} \right)^2 - 119.24 \frac{v_{Iso}}{v_{Iso} + v_{Hep}} \cdot \frac{v_{Tol}}{v_{Iso} + v_{Hep} + v_{Tol}} \right] + \quad \text{Eq. 4.12}$$

$$X_{Eth} MON_{Eth}$$

MON	motor octane number of the surrogate [-]
MON_{Eth}	motor octane number of ethanol [-]

The reaction kinetics mechanism used in this work is a further development of the mechanism presented in [26] and consists of 489 chemical species and 3369 elementary reactions [28]. The reference mechanism [26] includes the oxidation chemistry of various C_0–C_8 hydrocarbon species and substituted aromatic species, including n-heptane, iso-octane, toluene, and ethanol, allowing the formulation of multi-component surrogate fuels with capability of predicting Polycyclic Aromatic Hydrocarbon (PAH) formation in gasoline engines [54]. However, recent studies [29] have suggested the modification of the oxidation chemistry of alkanes in terms of thermochemistry, reaction rates, and pathways. The aim is to avoid the error compensation in chemical models as well as to improve the model's ability to predict ignition delay times accurately, which is of particular importance for modeling knock.

Therefore, the sub-models for n-heptane and iso-octane in the reference mechanism have been revised according to the state-to-art kinetic knowledge [28]. To this end, the combustion chemistry of n-heptane has been extracted from a recently published mechanism of normal alkanes [29]. Following the model development concept introduced in the same publication, a revised iso-octane mechanism has been developed in [28]. In order to optimize the computational performance, the detailed mechanisms for n-heptane and iso-octane have then been reduced to a skeletal level using a multi-stage reduction strategy proposed in [120]. Finally, the two reduced models have been built as additional modules upon the reference model [28].

4.3   Reaction Kinetic Simulations at In-Cylinder Conditions

The detailed kinetic reaction mechanisms presented in Section 2.2 cannot be integrated in 0D/1D simulations, as this would have a significant negative impact on the typically short computational times, Section 2.3.2. Hence, a simplified approach reproducing the behavior of the chemical processes in the unburnt mixture as accurately as possible is needed for modeling knock. To this end, typically the Livengood-Wu integral presented in Section 2.3.2.1 is used.

However, the analysis of measurement data in Section 3.3 confirmed that the commonly used knock integral is not capable of accurately estimating the progress of the chemical reactions leading to auto-ignition. To better understand how the chemical processes behave at in-cylinder conditions and to identify the flaws of the Livengood-Wu integral resulting in the observed poor performance of the commonly used simplified chemistry approach in Chapter 3, a kinetic simulation model was developed and implemented in MATLAB / Cantera, as described in Section 4.3.1. The model is scalable and can represent the unburnt zone of the Entrainment model presented in Section 2.3.1.2, or a small spot in it.

4.3.1   Simulation Model

The model for reaction kinetic simulations at in-cylinder conditions is based on an adiabatic reactor containing the surrogate-air-exhaust gas mass as shown in Figure 4.1. In addition, a moving wall was installed to compress and expand the mixture and hence reproduce the piston movement. The wall has heat transport properties that were used to recreate the wall heat losses calculated by the corresponding models within the performed pressure trace analysis of the measurement data, Section 3.2. As a mass flow from the unburnt into the burnt zone is used in the Entrainment model to represent the combustion, Section 2.3.1.2, the same technique was chosen for the simulation model in this work. The blowby losses estimated in the course of the performed PTA from a mass flow rate through an orifice resulting from a pressure gradient were considered too.

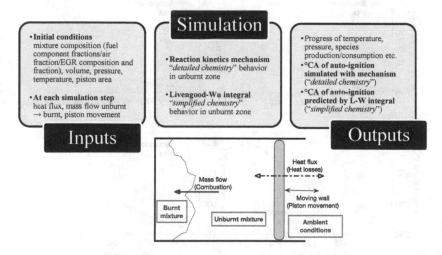

Figure 4.1:   Model for reaction kinetic simulations at in-cylinder conditions.

The theoretical air quantity required to complete the combustion of fuel can be calculated with the equation of stoichiometry of oxygen-fuel reactions, resulting in exactly $(m + n) / 4$ moles of oxygen for one mole of hydrocarbon C_mH_n. Together with the composition of air and the AFR value available in the measurement data, this reflection yields the fractions of air and fuel (sum of the four surrogate components) in the modeled cylinder.

Generally, the exhaust gas composition can be calculated from the surrogate component fractions and the AFR by assuming post-catalyst extraction and perfect catalytic conversion. However, even with these assumptions, rich combustion results in unburnt hydrocarbons (HC), carbon monoxide (CO), hydrogen (H_2) and nitric oxide (NO). For the case of excess fuel and recirculation of exhaust gas, the fractions of these species were estimated with empirical correlations derived from in-cylinder reaction kinetics simulations of single zone spark ignition combustion. The ignition of the mixture was implemented as an input of energy (heat flux) that was taken out of the system as soon as the mixture has ignited. The simulated behavior of the species of interest over various operating condition variations has been validated against both literature and measurement data, showing good agreement.

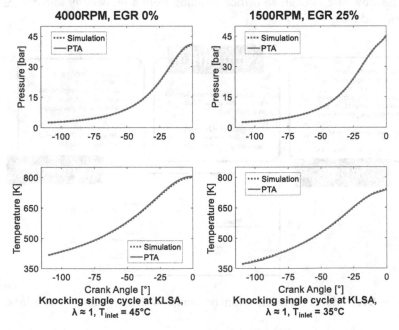

Figure 4.2:   Validation of the model for reaction kinetic simulations at in-cylinder conditions against PTA data.

The initial conditions needed at the start of the reaction kinetics simulations at in-cylinder conditions are the mixture composition (surrogate component fractions, AFR and EGR rate) as well as the initial pressure, temperature and volume (representing the piston position), Figure 4.1. The simulations yield if the mixture auto-ignites and when as well as the production and consumption of species causing changes in temperature and pressure. The model was validated by comparing simulated temperature profiles with PTA results during the compression stroke, where no heat release takes place. The curves show excellent agreement, Figure 4.2.

4.3.2 Investigation Findings

By simulating the measured knocking single cycles with the corresponding data from the preformed pressure trace analysis as described in Section 3.2, it was found out that the local auto-ignition in the unburnt mixture resulting in knock can occur in two stages. This phenomenon has already been observed in various studies in the context of gasoline Homogeneous Charge Compression Ignition (gHCCI) [119] [147], but has not been extensively investigated in the context of engine knock. The temperature profiles shown in Figure 4.3 illustrate the two-stage ignition behavior. The phenomenon is related to the negative temperature coefficient zone of gasoline fuels clearly visible in Figure 4.4, where the rate and extent of the hydrocarbon oxidation reactions, which at first increase rapidly with temperature, begin to fall with further rise in the temperature, causing an increase in the ignition delay times in this temperature region [44]. Further investigations relating the time of low-temperature (or the first stage of) ignition to the combustion progress have shown that the heat released during combustion always overlaps the low-temperature heat release. Thus, the first ignition stage cannot be identified in the heat release rate estimated by analyzing the combustion process.

Obviously, the amount of heat released during the low-temperature ignition depends on the operating conditions, e.g. engine speed and EGR rate as shown in Figure 4.3. Increasing engine speeds cause the temperature rise resulting from the low-temperature ignition to decline and high EGR rates lead to a later occurrence of the first ignition stage as well as a shorter time interval between the two ignition stages.

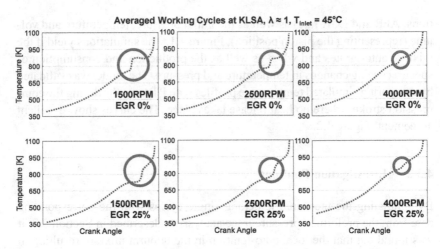

Figure 4.3: Simulated temperature profiles of single cycles with auto-ignition in two stages depending on the boundary conditions.

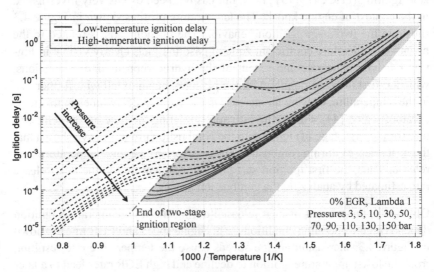

Figure 4.4: Two-stage ignition region and ignition delay times of the low- and high-temperature regimes (base gasoline surrogate).

Generally, two-stage ignition behavior characterizes iso-octane and n-heptane [147]. As these two hydrocarbons, together with toluene and ethanol, comprise the surrogate fuel, they also cause the observed low-temperature heat release. The resulting temperature increase influences the following chemical reactions and thus the ignition delay of the mixture significantly [147].

The boundary condition region where two-stage ignition of the base fuel-air mixture, Section 4.2, investigated in this study occurs is illustrated in Figure 4.4. The existence of an ignition delay time for the low-temperature ignition primarily depends on temperature and pressure and defines the two-stage ignition region of a given surrogate composition. On the other hand, the influence of EGR and the AFR on the region's size can be neglected. This fact is not to be mistaken with the effect of these parameters on the ignition delay of the mixture, which can be significant. Furthermore, Figure 4.4 clearly shows that, at high temperatures, the two-stage ignition phenomenon does not occur at all. This observation is relevant for both knock and gHCCI operation with high internal EGR rates, as these lead to high temperatures of the unburnt mixture [85]. In-cylinder reaction kinetics simulations of measured gHCCI cycles have confirmed that high internal EGR rates lead to unburnt temperature levels where no low-temperature heat release occurs, Figure 4.5, resulting in a single stage auto-ignition of the mixture.

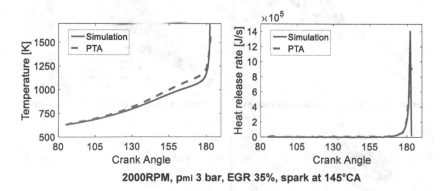

2000RPM, pmi 3 bar, EGR 35%, spark at 145°CA

Figure 4.5: Kinetic simulation of a measured gHCCI working cycle at a high internal EGR rate. The high temperature of the exhaust gas suppresses the occurrence of low-temperature heat release.

4.3.3 Auto-Ignition Prediction Performance of the Livengood-Wu integral

The two-stage auto-ignition event of the unburnt air-fuel mixture has a significant influence on the ignition delay. Hence, the question arises, if the commonly used Livengood-Wu integral can consider this effect and thus predict the time of auto-ignition accurately in case the mixture ignites in two-stages. Figure 4.6 shows a comparison between the crank angles of mixture auto-ignition and thus knock onset of the investigated single engine cycles simulated with the detailed mechanism at in-cylinder conditions (*"detailed / real chemistry"*, Section 4.3.1) and those predicted by the knock integral in Equation 2.11 (*"simplified chemistry"*) at various operating conditions. The ignition delay times needed at each integration step are obtained with the enhanced three-domain approach presented in Section 4.4.3, which accounts for the effects of all boundary conditions of relevance and reproduces the ignition delay times calculated with the detail reaction kinetics mechanism very accurately.

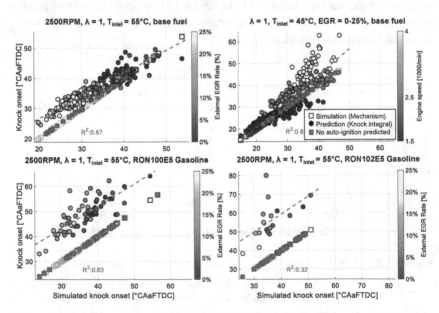

Figure 4.6: Times of auto-ignition (in °CA) of single cycles simulated with the detailed mechanism and predicted by a knock integral at various operating conditions.

Obviously, the occurrence of two-stage ignition results in partially huge errors and hence poor performance of the auto-ignition prediction performed with the commonly used Livengood-Wu integral. The prediction error varies significantly, reaching values of well over 20 °CA. Besides, an influence of engine speed, exhaust gas and surrogate composition on the behavior of the prediction quality is clearly visible. For some working cycles, no auto-ignition was predicted by the simplified approach at all, although the detailed mechanism auto-ignited, which represents the worst-case scenario. Thus, **it is obvious that the knock integral cannot account for the low-temperature ignition and its influence on the mixture's ignition delay. Livengood and Wu already surmised this circumstance in the course of their pioneer research in 1955 [105].**

The results from the performed investigations clearly show that the simplified modeling approach commonly used for predicting knock in 0D/1D SI engine simulations is not capable of reproducing the behavior of the detailed reaction kinetics mechanism at in-cylinder conditions. Hence, it is not a suitable replacement for the "detailed chemistry" and cannot be used for a reliable knock boundary prediction in SI engines running on gasoline. Thus, none of the available knock models for gasoline fuels based on the Livengood-Wu integral is fully predictive. Even if a knock model incorporates a very accurate sub-model for the ignition delay times of different mixtures at various boundary conditions derived from reaction kinetic simulations, such as the models proposed in [69] and [154], the reliable prediction of the knock boundary will not be possible without the repeated recalibration of the model at different operating conditions. **The general knock modeling approach has to be changed or improved, in order to consider the occurrence of two-stage ignition in the unburnt mixture that results in knock.** In this context, it should be remarked that for fuels that do not show a considerable two-stage ignition behavior such as methane, the Livengood-Wu integral has been proven to predict the time of auto-ignition in the unburnt mixture accurately [151].

4.4 Two-Stage Ignition Modeling

Although detailed reaction kinetic mechanisms are not suitable for direct use within a 0D/1D environment for computational time reasons, in the context of modeling knock, these can be employed to calculate the ignition delay times of air-fuel mixtures at various boundary conditions. Generally, the estimation of the empirical constants of the Arrhenius-type ignition delay models needed for the knock integral evaluation, e.g. Equation 2.12, is performed by fitting the knock integral and thus the underlying ignition delay equations to measurement data from an engine test bench by using the least squares method. Alternatively, reaction kinetic simulation results can be used to calibrate the Arrhenius-type models. Furthermore, they enable the development of advanced ignition delay modeling approaches that consider the effects of various boundary conditions and are appropriate for all temperature ranges of interest in an internal combustion engine, which is typically not the case with basic Arrhenius-type equations, as discussed in Section 2.3.2.1.

In this context, the ignition delay calculation with reaction kinetic mechanisms has the main advantage of being able to cover a much wider range of boundary conditions compared to the Arrhenius-type equation calibration based on measurement data from an engine test bench. Furthermore, values obtained using reaction kinetics are independent of engine-specific effects. Of course, these advantages are of relevance only if the mechanism used has been validated for all boundary conditions of interest.

Generally, the accuracy of reaction kinetic mechanisms is assessed based on literature data as well as ignition delay times measured with different apparatus, typically rapid compression machines, and shock tubes [26], which are discussed in detail in Section 2.2. However, the two appliances have fundamentally different working principles, so that facility effects of both setups have to be taken into account, which results in dissimilar measurement error tolerances (RCM typically 10 % and in case of a ST – up to 20 % [24]). Moreover, the magnitudes of the measured ignition delay times vary significantly (microseconds to seconds, see Figure 4.4). Shock tubes are typically used in the high temperature range, where the ignition delay is very short and thus cannot be determined in a rapid compression machine. In this context, it has to be remarked that the measurement of ignition delay times at high pressures is particularly difficult. Hence, the values simulated with a mechanism in the high-pressure region typically result from chemical reaction rates calibrated based on experiments at low pressures [158].

4.4.1   Simulation of Ignition Delay Times

With their reduced kinetic model, Hu and Keck [80] have successfully corre-
lated measurements of explosion limits in a constant volume bomb and ignition delay times in rapid compression machines. Consequently, Chun and Heywood [37] demonstrated the ability of this same kinetic model to predict the
onset of knock in a spark ignition engine. Thus, it has been revealed that the
explosion limits in a constant volume bomb, the ignition delay times in rapid
compression machines, and the prediction of knock onset are related. For this
reason, kinetic mechanisms are typically developed and validated based on
ignition delay times measured in rapid compression machines.

Similarly, the Livengood-Wu integral in Equation 2.11 aims to predict the time
of auto-ignition in SI engines based on the ignition delay times of the mixture
at a constant physical state. The formulation assumes a global reaction for the
concentration of chain carriers that reaches a critical value at the instant of
auto-ignition, Section 2.3.2.1. The auto-ignition process in rapid compression
machines is at constant temperature and pressure and thus physical state, as
the heat released in the course of the ignition process (during the ignition delay
interval) is minimal [119]. Hence, in this case the reaction rate can be related
directly to the inverse of the ignition delay time corresponding to the particular
temperature and pressure. Under engine conditions with varying temperatures
and pressures, the Livengood-Wu formulation implies that the critical concentration of chain carriers is reached by summing its continuous increment at
each instantaneous thermodynamic state (defined by temperature and pressure) imposed by the engine operation, with the accumulation rate in each interval represented by the corresponding ignition delay time.

These considerations imply that the ignition delay times at constant state
needed for the evaluation of the Livengood-Wu integral and hence knock
boundary prediction have to be estimated at a constant volume. To this end, an
ideal gas reactor model as described in Section 4.1 representing a closed thermodynamic system can be used, where the process of mixture auto-ignition is
adiabatic and isochoric. The ignition delay time calculations were performed
at various boundary conditions with the detailed mechanism already used for
the reaction kinetic simulations at in-cylinder conditions in Section 4.3. The
respective surrogate composition as well as the fuel, air and exhaust gas fractions were estimated as discussed in Section 4.3.1. However, the presence of
reactive species in the case of EGR and fuel excess, such as hydrogen (H_2) and

most importantly nitric oxide (NO), was deliberately neglected, as these have been reported to have an influence on knock in general as well as on the ignition delay of mixtures [149], and here the sheer dilution effect of EGR is of interest.

4.4.2 Definitions

An approach for modeling knock that accounts for a low-temperature ignition possibly taking place requires the development of suitable models for the three main parameters characterizing the two-stage ignition phenomenon. These are the ignition delay times of both ignition stages as well as the temperature increase resulting from the low-temperature ignition. First, the quantification of these three parameters is necessary. To this end, following definitions are introduced, as shown in Figure 4.7:

■ The location of the low-temperature ignition and hence the low-temperature ignition delay τ_{low} are defined as the point max(T_{grad}), where the temperature gradient reaches its maximum before the auto-ignition of the mixture.

■ The high-temperature or the auto-ignition delay τ_{high} is detected based on a threshold for the temperature gradient. This method has already been suggested in other studies [137]. The value of the temperature gradient threshold was set to 25 K per microsecond.

■ The temperature increase T_{incr} resulting from the low-temperature ignition is defined as temperature at simulation start subtracted from the temperature at the point min(T_{grad}), where the temperature gradient reaches its minimum between the two ignition stages. This definition has already been suggested in previous studies too [119].

The occurrence of two-stage ignition can be detected based on chain branching reactions producing OH radicals, Section 2.1.2, followed by oxidation reactions that consume the produced radicals and result in heat release and consequentially temperature increase occurring before the mixture's auto-ignition, Figure 4.7. Additionally, a variable simulation step was implemented to save computational time.

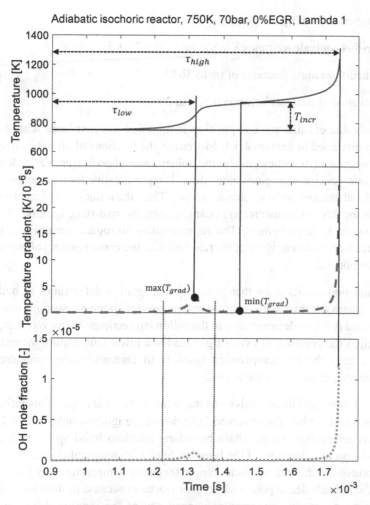

Figure 4.7: Definition of low- and high-temperature ignition as well as temperature increase resulting from the first ignition stage in an adiabatic isochoric reactor.

The boundary conditions were varied according to in-cylinder values typical for SI engines today with the main goal being to cover a sensible value range as wide as possible:

■ Pressures of up to 150 bar;

■ Temperatures of up to 1300 K;

■ Air-fuel equivalence ratios λ in the interval 0.7 - 1.5;

■ Exhaust gas mass fractions of up to 30 %;

■ Fuel ethanol contents between 0 % and 20 %.

Obviously, the exhaust gas composition is modified in accordance with the AFR, as discussed in Section 4.3.1. Moreover, the fractions of all base surrogate components (iso-octane, n-heptane, toluene and ethanol) were varied too, as only one, specific surrogate composition is representative for a given gasoline fuel and matches up to its characteristics. Thus, the influence of fuel properties on the three main parameters characterizing the two-stage ignition phenomenon can be apprehended. The representative surrogate composition is calculated from characteristics of the real fuel with the corresponding blending rules, Section 4.2.

The simulation results show that generally, the ignition delay times of both ignition stages decrease with pressure and increase with the exhaust gas mass fraction and AFR, as demonstrated in the following sections. Moreover, rising λ as well as the presence of exhaust gas lead to a lower temperature increase resulting from the low-temperature ignition. In contrast, higher pressures cause the temperature increase to rise.

A detailed chemical kinetic analysis of the ignition at a wide range of operating conditions can explain the observed behavior of the ignition delay times. In the lower temperature range, chain branching reactions build up the radical pool and lead to a decrease of the ignition delay with temperature. At higher temperatures in the intermediate range (negative temperature coefficient zone), the radicals decompose back to their reactants because of their instability, causing a negative temperature dependency of the ignition delay time [158]. At temperatures higher than the intermediate regime, the high-temperature reactions dominate, resulting in typically short ignition delay times. Furthermore, the complex dependence of the chemical reactions on pressure is not as strongly pronounced as the influence of temperature. Detailed information on the reaction pathways, rate rules and thermochemistry leading to the effects of the investigated boundary condition variations on the mixture ignition delay is available in numerous publications, for example [118] [150] [158]. As these are not the focus of this work, the chemical kinetic background will not be further discussed here.

The obtained simulation data were used to develop new models for the calculation of the three main parameters characterizing the two-stage ignition phenomenon, with the two main goals being the achievement of high accuracy of the calculated values and short computational times. Clearly, the NTC zone has to be modeled as precisely as possible because of its significant influence on the auto-ignition delay times, Figure 4.4. All estimated model parameters should only be recalibrated if values of the parameters characterizing the two-stage ignition phenomenon obtained with a reaction kinetics mechanism different than the one used in this work have to be considered. The developed models can afterwards be installed in a new knock modeling approach that accounts for the possible occurrence of two-stage ignition and its influence on the mixture's ignition delay.

4.4.3 High-Temperature Ignition Delay

The negative temperature coefficient zone typical for common gasoline fuels has a significant influence on the auto-ignition delay times in the intermediate temperature range. In this context, over 60 years ago Schmidt and Beckers remarked that in case of gasoline fuels, Arrhenius-type ignition delay models are only adequate in a limited range of temperature for any particular pressure, Section 2.3.2.1. This fact also becomes apparent from the well-known, commonly used logarithmic ignition delay plots, where the exponential Arrhenius-type equations are represented by straight lines as shown in Figure 4.8. In contrast, the ignition delay curves of the selected base fuel estimated with the detailed reaction kinetics mechanism have a much more complex shape.

Hence, it is clear that simple Arrhenius-type equations are not appropriate for modeling the auto-ignition delay times of gasoline fuels, as their simplicity leads to huge errors in the empirically calculated results. Consequently, as the majority of the commonly used 0D/1D knock models incorporate simple Arrhenius-type ignition delay models, Section 2.3.2.2, these do not consider the fuel's negative temperature coefficient behavior and perform knock predictions based on hugely inaccurate ignition delay time values, which is demonstrated in Figure 4.8.

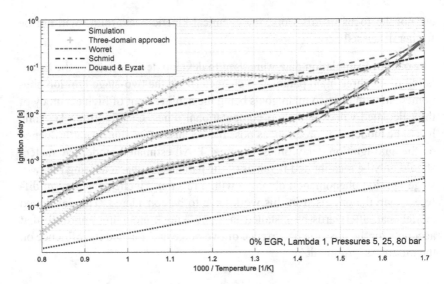

Figure 4.8: Comparison of the auto-ignition delay models incorporated into commonly used knock models with values obtained from simulations with a detailed reaction kinetics mechanism.

To this end, Weisser [160] proposed a distinction of low-, medium- and high-temperature regimes of ignition, Figure 4.9, resulting in the 3-domain approach for the calculation of the auto-ignition delay times of air-fuel mixtures in Equation 4.13. $\tau_{1,high}$, $\tau_{2,high}$ and $\tau_{3,high}$ represent the ignition delays of the mixture in the low-, medium- and high-temperature regimes of ignition respectively. This ignition delay modeling approach, originally developed for n-heptane-air mixtures, is particularly suitable for fuels with distinctive negative temperature coefficient behavior, as is the case with gasoline.

Each of the three timescales in Equation 4.13 can be evaluated using a single Arrhenius-type correlation as shown Equation 4.14. Generally, the Arrhenius-type equations used for the calculation of the mixture's ignition delay are expanded by adding parameters in front of the exponential function to account for the influence of various changes in the boundary conditions. In the meantime, the value of the model parameter in the numerator of the exponential function, which is proportional to the activation energy of the global chemical reaction leading to auto-ignition, is kept constant. By keeping the numerator of the exponential function constant, the Arrhenius-type equation can easily be fitted to the ignition delay data. This approach leads to an auto-ignition model error though, which in turn has to be compensated for by the coefficients in front of the exponential function, as in reality the activation energy changes significantly with the boundary conditions [137].

$$\frac{1}{\tau_{high}} = \frac{1}{\tau_{1,high} + \tau_{2,high}} + \frac{1}{\tau_{3,high}}$$
Eq. 4.13

τ_{high}	high-temperature (auto-) ignition delay [s]
$\tau_{1,high}$	high-temperature ignition delay in low-temperature regime of ignition [s]
$\tau_{2,high}$	high-temperature ignition delay in medium-temperature regime of ignition [s]
$\tau_{3,high}$	high-temperature ignition delay in high-temperature regime of ignition [s]

$$\tau_{i,high} = A_{i,high} \cdot e^{\left(\frac{B_{i,high}}{T}\right)}$$
Eq. 4.14

$$A_{i,high}, B_{i,high} = F_{1,2}(p, \lambda, EGR, surr.\,comp.\,fracs.)$$

i	temperature regime index [-]
$A_{i,high}$	pre-exponential factor, high-temperature ignition delay [-]
$B_{i,high}$	activation energy parameter, high-temperature ignition delay [K]
$F_{1,2}$	empirical functions

However, it is much more convenient to assemble all influences of the boundary conditions in two parameters $A_{i,high}$ and $B_{i,high}$, with i denoting the temperature regime, resulting in the simple formulation shown in Equation 4.14. In this case, the model parameter inside the exponential function $B_{i,high}$ is also a function of the boundary conditions. Thus, boundary condition changes are correctly reflected in different values of the activation energy. Furthermore, thanks to the additional degree of freedom in Equation 4.14 (the influence of each parameter is considered in front of and inside of the exponential function), the approach can be fitted to much more complexly shaped ignition delay curves. This yields an enhanced, flexible, and very powerful ignition delay modeling approach. Disadvantageous is that, partially because of its flexibility, the approach proposed here is more difficult to fit to the available ignition delay data.

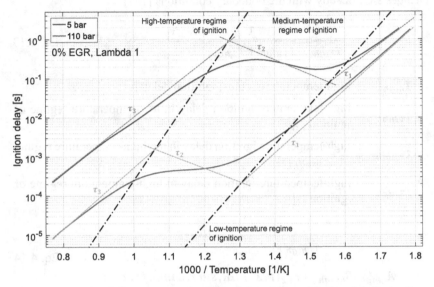

Figure 4.9:   Modeling the auto-ignition delay times of fuels with distinctive negative temperature coefficient behavior with a three-domain approach.

In order to overcome the problems resulting from the complexity of the proposed ignition delay modeling approach, the estimation of the model parameters in Equation 4.14 can be fully automated, yielding good model accuracy as shown in left plot in Figure 4.10. However, the achieved quality of the calculated ignition delay times is still not satisfactory, especially if the logarithmic scale of the y-axis is considered. Moreover, the most significant model errors occur in the regions where the ignition delay times are small and thus the integrand $1 / \tau$ substantial, resulting in a rapid increase of the knock integral value.

To this end, an automatic three-stage optimization of the estimated model parameters was implemented by using the algorithm described in [95] to minimize the normalized root mean square deviation of the empirically calculated ignition delay times. The first optimization stage involves the error minimization for each of the simulated ignition delay curves, and is followed by the reduction of the total model deviation over all simulated ignition delay curves. The implemented normalization accounts for the different magnitudes of the values of the ignition delay times. Finally, the interpolation and extrapolation capabilities of the model are verified with simulation data that has not been used in the course of the ignition delay model optimization.

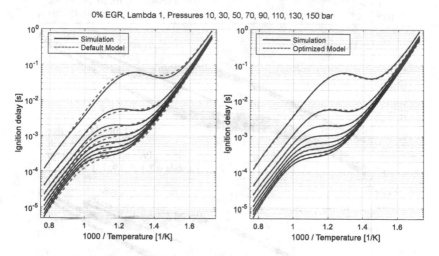

Figure 4.10: Accuracy gain achieved by automatically optimizing the estimated high-temperature ignition delay model coefficients.

The enhanced 3-domain approach, together with the developed methods for automating and optimizing the model parameters, result in a new high-temperature ignition delay model that yields results with very high accuracy, as shown in Figure 4.11 and Figure 4.12. It should be remarked that the deterioration of the model accuracy in the NTC zone at pressures below 5 bar is not relevant for the prediction of knock in SI engines, because the occurrence of such high temperatures at pressures that are so low is not realistic in an internal combustion engine.

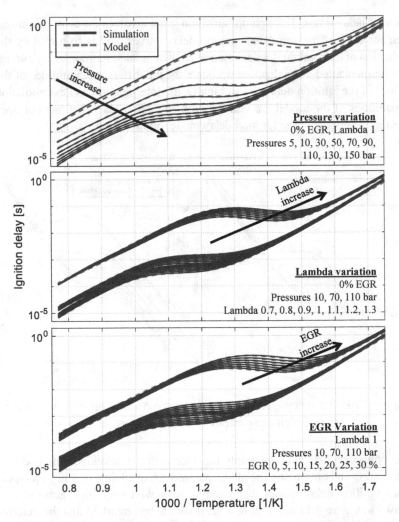

Figure 4.11: Simulated and modeled high-temperature ignition delay times of the base gasoline surrogate at various boundary conditions.

Overall, the newly developed high-temperature ignition delay model reproduces exactly the negative temperature coefficient behavior typical for gasoline fuels and correctly considers the influences of mixture dilution with air and / or exhaust gas and / or water (Section 4.5.4), as well as the fuel properties (resulting in different surrogate component fractions and fraction ratios, Section 6.2) at various temperatures and pressures. The model equations for the parameters $A_{i,high}$ and $B_{i,high}$ as well as the estimated values of the model coefficients are listed in the Appendix, Section A1.

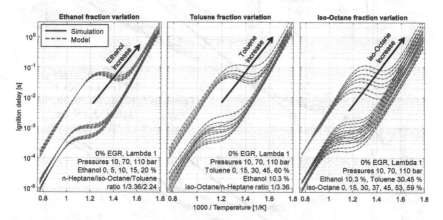

Figure 4.12: Simulated and modeled high-temperature ignition delay times of different surrogate compositions at various boundary conditions.

4.4.4 Low-Temperature Ignition Delay

For the low-temperature ignition delay τ_{low}, it is convenient to choose a modeling approach similar to the one for the auto-ignition delay times. As discussed in Section 4.3.2, the existence of an ignition delay time for the low-temperature ignition primarily depends on temperature and pressure and defines the two-stage ignition region of a given surrogate composition. Furthermore, Figure 4.13 and Figure 4.14 clearly show that at high temperatures, the two-stage ignition phenomenon does not occur at all. Thus, the low-temperature ignition delay curves are characterized by just two temperature regimes of ignition. Hence, the modeling approach has to account for the low and medium regimes only as shown in Equation 4.15.

As is the case in Equation 4.14, the Arrhenius-type correlations for the two low-temperature ignition temperature regimes include two parameters $A_{i,low}$ and $B_{i,low}$ that change with the boundary conditions as shown in Equation 4.16.

Thus, like in the previous section, any alternations in the boundary conditions are reflected in an activation energy change.

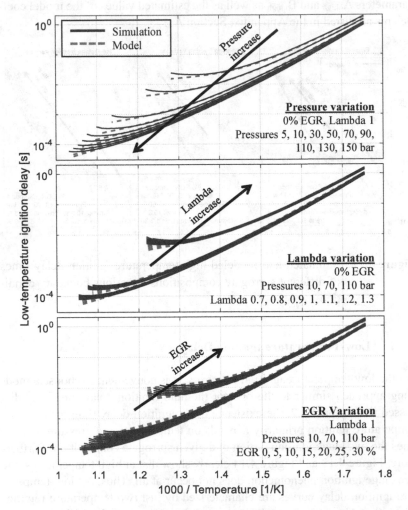

Figure 4.13: Simulated and modeled low-temperature ignition delay times of the base gasoline surrogate at various boundary conditions.

The parameter estimation and optimization techniques used in the low-temperature ignition delay model do not differ from those already discussed in the previous section. Thus, high accuracy of the calculation results could be

achieved, as shown in Figure 4.13 and Figure 4.14. The slight differences between the simulation results and the values calculated with the new model that
can be observed near the boundary of the two-stage ignition region at low
pressures do not influence the quality of the two-stage ignition prediction accuracy, as will be demonstrated in the course of the auto-ignition approach
validation in Section 4.5.3.

$$\frac{1}{\tau_{low}} = \frac{1}{\tau_{1,low} + \tau_{2,low}}$$

Eq. 4.15

τ_{low} low-temperature ignition delay [s]

$\tau_{1,low}$ low-temperature ignition delay in low-temperature regime of ignition [s]

$\tau_{2,low}$ low-temperature ignition delay in medium-temperature regime of ignition [s]

$$\tau_{i,low} = A_{i,low} \cdot e^{\left(\frac{B_{i,low}}{T}\right)}$$

Eq. 4.16

$$A_{i,low}, B_{i,low} = f_{1,2}(p, \lambda, EGR, surr. comp. fracs.)$$

$A_{i,low}$ pre-exponential factor, low-temperature ignition delay [-]

$B_{i,low}$ activation energy parameter, low-temperature ignition delay [K]

$f_{1,2}$ empirical functions

The low-temperature ignition delay model proposed here correctly reproduces
the influences of mixture dilution with air and / or exhaust gas and / or water
(Section 4.5.4), as well as the fuel properties (resulting in different surrogate
component fractions and fraction ratios, Section 6.2) at various temperatures
and pressures. The model equations for the parameters $A_{i,low}$ and $B_{i,low}$ as well
as the estimated values of the model coefficients are listed in the Appendix,
Section A2.

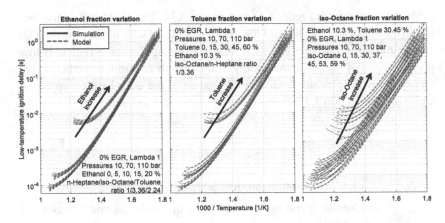

Figure 4.14: Simulated and modeled low-temperature ignition delay times of different surrogate compositions at various boundary conditions.

4.4.5　Temperature Increase Resulting from Low-Temperature Ignition

The occurrence of low-temperature ignition results in heat release, thus leading to a temperature increase. Generally, it is possible to model each of these two parameters. However, because the temperature is an input parameter of the auto-ignition delay model in Equation 4.14, it is more convenient to model the temperature increase instead of the heat released during the first ignition stage.

There is one temperature increase value corresponding to each existing low-temperature ignition delay. As the occurrence of two-stage ignition and thus the presence of a low-temperature ignition delay value are primarily influenced by temperature and pressure, as discussed in Section 4.3.2, these two parameters are also decisive for the existence of a value for the temperature increase at the current boundary conditions. Furthermore, exhaust gas and the AFR lead to significant changes in the magnitude of low-temperature heat release, Figure 4.15.

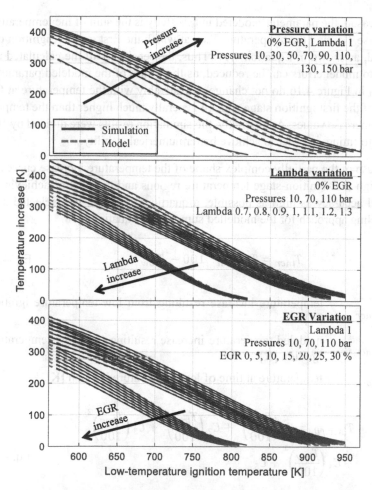

Figure 4.15: Simulated and modeled temperature increase resulting from low-temperature ignition of the base gasoline surrogate at various boundary conditions.

The temperature is by far the most critical boundary condition in respect of the auto-ignition behavior of air-fuel mixtures – a fact that is represented by the exponential influence of this parameter on the ignition delay times of both ignition stages, see Equation 4.14 and Equation 4.16. Hence, the ignition delay times are very sensitive to changes in temperature and thus errors in the calculated temperature increase resulting from the low-temperature heat release. For this reason, the parameter modeled in this study is the sum of the temperature increase T_{incr} and the temperature T_{low}, at which the first stage of ignition occurred, as shown in Equation 4.17. Thus, the sensitivity of the calculated results to model errors can be reduced, as the values of the modeled parameter $T_{incr,fit}$ in Figure 4.16 do not change considerably with the temperature at the time of the first ignition stage and are generally much higher than the temperature increase values. Additionally, the simulation results were divided by 100 prior to fitting, in order to achieve better numerical stability.

Because of the partially complex shape of the temperature increase curves in the high first-ignition-stage temperature regions and the goal of achieving a model accuracy as high as possible, a quartic polynomial was selected as a modeling approach for the modelled sum $T_{incr,fit}$, Equation 4.18.

$$T_{incr} = T_{incr,fit} \cdot 100 - T_{low} \qquad \text{Eq. 4.17}$$

T_{incr} temperature increase resulting from low-temperature ignition [K]

$T_{incr,fit}$ modeled temperature increase resulting from low-temperature ignition [K]

T_{low} temperature at time of low-temperature ignition [K]

$$T_{incr,fit} = C_1 \left(\frac{T_{low}}{100}\right)^4 + C_2 \left(\frac{T_{low}}{100}\right)^3 + C_3 \left(\frac{T_{low}}{100}\right)^2$$
$$+ C_4 \left(\frac{T_{low}}{100}\right)^1 + C_5 \qquad \text{Eq. 4.18}$$

$$C_{1,2,3,4,5} = g_{1,2,3,4,5}(p, \lambda, EGR, surr. comp. fracs.)$$

$C_{1,2,3,4,5}$ temperature increase model parameters [$1/K^3$, $1/K^2$, $1/K$, -, K]

$g_{1,2,3,4,5}$ empirical functions

The parameter estimation and optimization techniques used in the temperature increase model do not differ from those already discussed in the high- and low-temperature ignition delay sections. Thus, the temperature increase model yields results with very high accuracy, Figure 4.15 and Figure 4.17, and correctly reproduces the influences of mixture dilution with air and / or exhaust gas and / or water (Section 4.5.4), as well as the fuel properties (resulting in different surrogate component fractions and fraction ratios, Section 6.2) at various temperatures and pressures. The model equations for the parameters $C_{1..5}$ as well as the estimated values of the model coefficients are listed in the Appendix, Section A3.

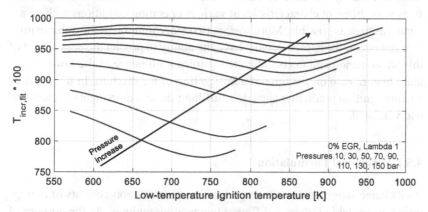

Figure 4.16: Simulated influence of temperature and pressure on the modeled sum of temperature increase and temperature, at which the first stage of ignition occurred.

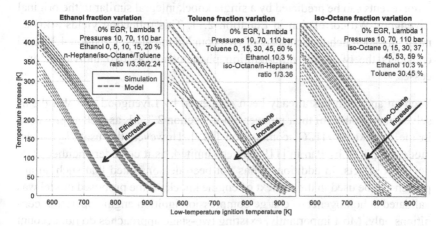

Figure 4.17: Simulated and modeled temperature increase resulting from low-temperature ignition of different surrogate compositions at various boundary conditions.

4.5   Two-Stage Auto-Ignition Prediction Approach

A knock model that accounts for the possible occurrence of low-temperature ignition has to incorporate an auto-ignition prediction approach capable of reliably estimating the time of auto-ignition of air-fuel mixtures for gasoline fuels with different characteristics at various operating conditions with an accuracy as high as possible. Moreover, as two-stage ignition of the unburnt mixture can occur, the approach must have the ability to predict the appearance of this phenomenon and to consider the significant influence of the low-temperature heat release on the time of auto-ignition of the mixture. In this section, an auto-ignition prediction approach fulfilling these requirements is proposed and validated.

4.5.1   General Formulation

As its name suggests, the two-stage ignition phenomenon consists of two ignition events taking place at different temperature regimes. As the purpose of the Livengood-Wu integral is to predict if and when a pre-defined, critical pre-reaction state of the unburnt mixture will be exceeded, resulting in an auto-ignition in the end-gas, Section 2.3.2.1, the occurrence of each of the two ignition events can be predicted by a single knock integral similar to the original Livengood-Wu correlation in Equation 2.11. This reflection yields a knock modeling approach that consists of two coupled integrals, Equation 4.19, and considers the two-stage auto-ignition phenomenon as two individual processes.

Such an approach has already been suggested by Livengood and Wu themselves in the course of their pioneer work, Section 2.3.2.1, as well as proposed in the context of gHCCI combustion [119]. However, the usability of the model proposed by Pan in [119] is very limited, as it considers neither EGR nor AFR effects. In addition, it has not been demonstrated that such an approach can be used in the context of engine knock, as the proposed model was validated at homogeneous charge compression ignition engine operation conditions only. Most importantly, existing two-stage approaches do not account for the progress of the aggregate auto-ignition reaction after the low-temperature ignition, leading to partially significant prediction errors [53]. As reported in [74], it has to be considered that after the low-temperature ignition occurrence, the aggregate auto-ignition reaction has progressed, as many chemical reactions have already taken place. Hence, the pre-reaction state of the mixture after the low-temperature ignition is not zero.

$$1 = \int_0^{t_1} \frac{dt}{\tau_{low}(BC)} \quad \xrightarrow[\substack{\text{Progress of aggregate} \\ \text{reaction (0.3)}}]{\substack{\text{Temperature increase } T_{incr} \\ \text{Pressure increase } p_{incr}}} \quad 1 = \int_{t_1}^{t_2} \frac{dt}{\tau_{high}(BC, T_{incr}(BC), p_{incr}(T_{incr}))} \qquad \text{Eq. 4.19}$$

t_1 predicted time of low-temperature ignition [s]

t_2 predicted time of high-temperature (auto-) ignition [s]

p_{incr} pressure increase resulting from low-temperature ignition [bar]

BC boundary conditions

The inputs of the two coupled integrals are the values of the ignition delay for the corresponding ignition stage as a function of the current boundary conditions, denoted by BC in Equation 4.19. These are calculated by the corresponding models presented in Sections 4.4.3, 4.4.4 and 4.4.5. As in the original Livengood-Wu correlation in Equation 2.11, the end-of-integration value of both ignition stages is one.

The time of low-temperature ignition t_1 is both upper limit of the first integral and bottom limit of the second integral, meaning the second integral representing the high-temperature ignition starts as soon as the first integral has reached one and the low-temperature ignition has occurred. Consequently, the temperature increase T_{incr} resulting from the first ignition stage, as well as the corresponding pressure increase p_{incr}, are evaluated at the predicted time of low-temperature ignition t_1 and have to be added to the respective unburnt mixture values at each integration step of the second integral, as shown in Figure 4.18. Thus, the influence of the low-temperature ignition on the auto-ignition delay of the mixture is considered. Mixture auto-ignition is defined as the point where the high-temperature ignition integral reaches one – a value that is independent of the operating conditions and corresponds to the constant pre-reaction state of the mixture that has to be reached and exceed for auto-ignition to occur. The integration is performed in the time domain with a small, constant step size that was set to 10^{-7} seconds in this work.

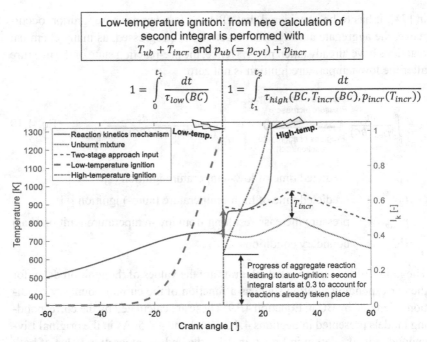

Figure 4.18: Auto-ignition prediction for one engine cycle with the newly developed two-stage approach.

The new two-stage auto-ignition prediction approach requires the very accurate calculation of the ignition delay times of both ignition stages as well as the temperature increase, especially because an inaccurate prediction of the low-temperature ignition point t_1 means the second integral will start too early or too late, leading to an error in the prediction of the auto-ignition point t_2. Furthermore, in this case the temperature increase caused by the low-temperature ignition will not be calculated at the correct boundary conditions. This will result in an error in the most important input of the second integral (exponential temperature dependence of the ignition delay in Equation 4.14 and hence in the predicted time of mixture's auto-ignition t_2.

As already discussed, besides the temperature increase, in case of a two-stage ignition the progress of the aggregate auto-ignition reaction after the first ignition stage has to be considered and hence modeled too [74]. Thus, it is accounted for the fact that at the starting point of the high-temperature reactions (and the corresponding integral in Equation 4.19), the pre-reaction state of the unburnt mixture is not zero. This is because chemical reactions have already taken place and many intermediate species have been produced, which influences the following reactions significantly. This circumstance can also be observed in Figure 4.18: the reaction kinetics mechanism temperature almost does not change after the first ignition stage, although the cylinder volume is increasing (expansion stroke, FTDC is at 0 °CA), thus suggesting that chemical reactions are already taking place and some heat is being released before the auto-ignition. In contrast and as expected, the unburnt mixture temperature estimated with the two-zone SI combustion model decreases as the mixture is being expanded. As soon as both the reaction kinetics and 0D unburnt temperatures start to rise due to the combustion and hence the decrease of the unburnt mass (mass transfer from the unburnt zone into the burnt zone), the mixture in the kinetic simulation auto-ignites.

Because of these complex interrelations, it is clear that considering the influence of low-temperature ignition by simply adding the temperature increase T_{incr} to the 0D unburnt temperature values and thus shifting the 0D curve towards higher temperatures as shown in Figure 4.18 is not sufficient. Detailed investigations showed that such a simplified approach leads to partially huge errors in the predicted time of mixture auto-ignition that strongly depend on the operating conditions. This observation confirms the conclusions drawn from the investigations performed by Hernández in [74], where it is suggested that the simulation of the second auto-ignition stage requires the preliminary estimation of both the composition of the mixture and the temperature increase at the end of the first ignition stage. However, this means that intermediate species that cannot be obtained from universal equations have to be calculated. Hence, their estimation has to be performed with a reaction kinetics mechanism that on the other hand is not suitable for direct use within a 0D/1D environment because of the high computational effort it results in. Hence, the composition of the mixture after the onset of the cool flame (first ignition stage), which is largely dependent on the initial as well as the operating conditions, has to be estimated with a reaction kinetics mechanism in an adiabatic isochoric reactor at various boundary conditions and then modeled, as done for the three parameters characterizing the two-stage ignition phenomenon in Section 4.4.1. However, in the case of the mechanism used in this work, the mixture is comprises almost 500 different species whose concentrations (or fractions) have to be modeled as a function of the boundary conditions at the time of low-temperature ignition (pressure, exhaust gas fraction etc.), implying a very complex modeling approach and great effort for the estimation of its co-efficients. Besides, because of this complexity, the model quality and the accuracy of the calculation results is questionable.

Alternatively, [74] proposed to add a term to the original Livengood-Wu integral method to evaluate the influence of cool flames. The suggested equation includes a coefficient that estimates the weight of the low temperature chemistry over the overall process. This coefficient is obtained by using an optimization methodology. In the case of fuels burning with no cool flames, it is set to zero. In contrast, Pan [119] neglected the influence of the aggregate reaction progress after the low-temperature heat release completely, thus proposing a two-stage auto-ignition prediction approach for gHCCI combustion that is characterized by partially huge prediction errors. In this work, the progress of the aggregate auto-ignition reaction after the first ignition stage was taken into consideration by setting the starting value of the high-temperature ignition integral to 0.3, as shown in Figure 4.18. This value was estimated based on an evaluation of the thousands single cycles simulated with the detailed mechanism at various operating conditions and for different gasoline fuels. Furthermore, despite the huge number of parameters significantly influencing the auto-ignition behavior of air-fuel mixtures that have been varied in the course of the performed investigations, it has proven sufficient to keep the starting value of the second integral of 0.3 constant. Thus, the time of auto-ignition of air-fuel mixtures can be predicted very accurately at various operating conditions, as demonstrated in Section 4.5.3.

Hence, the pre-reaction state reached after the low-temperature ignition is assumed independent of the boundary conditions as well as the fuel composition. It equals 0.3, meaning that at the time of low-temperature ignition the progress of the global aggregate reaction leading to auto-ignition is always 30 %. Consequently, the two-stage knock modeling approach involves the calculation of two integrals for the low- and high-temperature ignition respectively, where the second integral starts at 0.3 and the temperature and pressure changes caused by the low-temperature ignition are added to the corresponding high-temperature integral inputs. Obviously, for the cycle shown in Figure 4.18, the time of auto-ignition could be predicted very accurately, as the locations where the detailed mechanism auto-ignites and the high-temperature ignition integral reaches one coincide.

4.5.2   Prediction of Two-Stage Ignition Occurrence

The low-temperature ignition integration process in Equation 4.19 can only take place as long as a low-temperature ignition delay exists for the boundary conditions at each integration step, Figure 4.19. This requirement is fulfilled inside the two-stage ignition region shown in Figure 4.4 and Figure 4.19. Consequently, if a point outside of this region is reached during the course of the low-temperature ignition integration process that is still running, as the integral has not reached one, a corresponding low-temperature ignition delay will not be available and further performing the integration will not be possible. Hence, in this case the low-temperature ignition region has been passed through without two-stage ignition occurring.

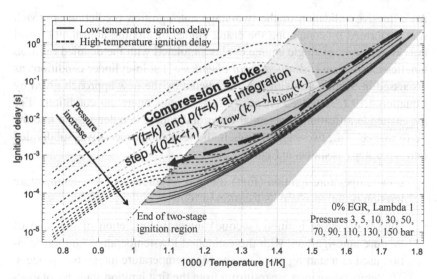

Figure 4.19: Prediction of low-temperature ignition occurrence for one engine cycle.

On these terms, the high-temperature integration process Equation 4.19 is directly started at the beginning of the model calculation (t = 0), as it is clear that in this case the auto-ignition results from chemical reactions in the high-temperature regime, and neither a low-temperature ignition integral nor a temperature increase are considered. Consequently, if no two-stage ignition occurs, the common single stage knock integral approach is used, as shown in Equation 4.20. Of course, in this case the ignition delay times are further calculated with the corresponding high-temperature ignition delay model proposed in Section 4.4.3.

$$1 = \int_0^{t_2} \frac{dt}{\tau_{high}(BC)}$$

Eq. 4.20

4.5.3 Approach Validation

An extensive validation of the two-stage auto-ignition prediction approach was performed by comparing the crank angles of mixture auto-ignition and thus knock onset of single engine cycles simulated with the detailed reaction kinetics mechanism (*"detailed / real chemistry"*) at in-cylinder conditions as discussed in Section 4.3 with those predicted by the new approach given by Equation 4.19 (*"simplified chemistry"*) at various operating conditions. For the easier identification of potential flaws and a more detailed insight into the performance of the two-stage approach, the validation process was subdivided into three steps as shown in Figure 4.20:

a. Low-temperature ignition (first) integral: Prediction of low-temperature ignition occurrence and its location in °CA;

b. High-temperature ignition (second) integral: Prediction of the time of mixture auto-ignition, with the estimated low-temperature ignition location used as a starting point for the high-temperature integration process. The temperature increase resulting from the first ignition stage is not evaluated;

c. Temperature increase and progress of aggregate reaction: Prediction of the temperature and pressure increase resulting from occurred low-temperature heat release as well as estimation of the progress of the aggregate reaction leading to auto-ignition after the first ignition stage (set to 30 % as discussed in the previous section) and investigation of their influence on the estimated time of mixture auto-ignition.

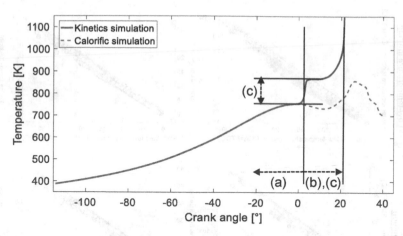

Figure 4.20: Kinetic and calorific unburnt temperature curves of one engine cycle for the two-stage approach validation.

Both validation steps a. and b. were performed with the temperature curves obtained from kinetic simulations performed with the detailed kinetic reaction mechanism at in-cylinder conditions as shown in Figure 4.20. Hence, these two steps exclude the prediction of the temperature increase caused by the low-temperature ignition, as its value is implicitly included in the kinetic simulations' temperature curves. The validation of the low-temperature heat release prediction itself was performed with temperature curves resulting from calorific simulations (no chemical reactions taking place) of the two-zone combustion model's unburnt zone with the detailed mechanism. These are almost identical with the 0D/1D unburnt temperature curves of the investigated single cycles. The minor differences result from the simplified calorific approaches used in 0D/1D engine simulations that consider only a small number of species [65] [68]. Results from all three validation steps a., b. and c. demonstrating the prediction capabilities of the two-stage knock modeling approach are shown in Figure 4.21 and Figure 4.22. Generally, the predicted times of both ignition stages match the corresponding points of auto-ignition calculated with the detailed mechanism very good within all operating condition variations performed.

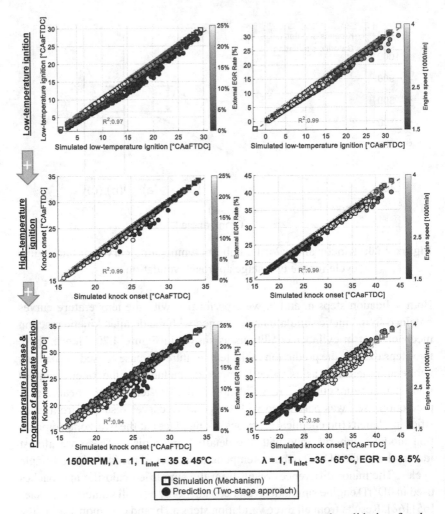

Figure 4.21: Two-stage auto-ignition prediction approach validation for the base fuel at various operating conditions, part 1.

The top and middle plots in Figure 4.21 and Figure 4.22 show the results from validation steps a. and b. respectively. Obviously, some of the errors in the calculated time of low-temperature ignition influence the respective starting points of the second integral, resulting in a marginal error in the predicted mixture auto-ignition location. Additionally, despite the exact low-temperature ignition prediction, the calculated time of auto-ignition of some cycles is slightly inaccurate. In both cases, the prediction errors are down to inaccuracies of the two models for the ignition delay times presented in Sections 4.4.3 and 4.4.4.

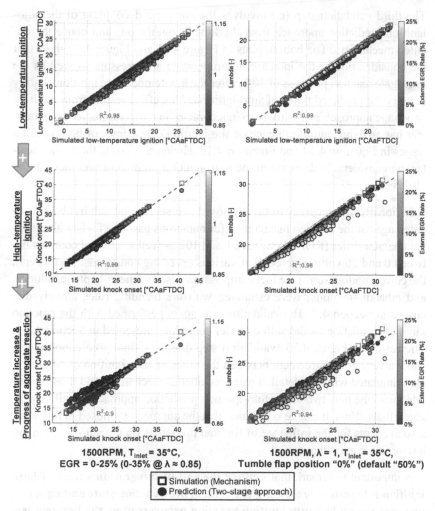

Figure 4.22: Two-stage auto-ignition prediction approach validation for the base fuel at various operating conditions, part 2.

Nevertheless, the overall calculation errors are very small. Hence, the prediction capabilities of the two-stage approach in Equation 4.19 coupled with the powerful models for the ignition delay times of both ignition stages can be quantified as very high.

The third validation step (c.) involves the complete decoupling of the auto-ignition prediction approach from the kinetic simulations and hence the detailed mechanism. The bottom plots in Figure 4.21 and Figure 4.22 show that the consideration of the increase in temperature and pressure as well as the aggregate reaction progress of 30 % after the low-temperature ignition slightly scatters the predicted times of auto-ignition. This effect results from the high two-stage approach's sensitivity to inaccuracies in temperature, as the parameter has an exponential influence on the ignition delay times of both ignition stages in Equation 4.14 and Equation 4.16. However, despite the higher scatter, no considerable decrease in the auto-ignition prediction accuracy can be observed.

Additionally, the accuracy of the proposed two-stage approach has been validated against the detailed mechanism for numerous gasoline fuels with different characteristics (RONs between 91 and 102 as well as ethanol contents between 0 and 20 volume percent) at various operating conditions. To this end, the corresponding surrogate fuel compositions (iso-octane, n-heptane, toluene, and ethanol fractions) were estimated with the blending rules already discussed in Section 4.2. The validation was again performed with the reaction kinetics simulation model at in-cylinder conditions presented in Section 4.3.1. The results in Figure 4.23 (validation step c. only – final prediction values) show a very good agreement between the times of auto-ignition of single cycles simulated with the detailed reaction kinetics mechanism and those calculated with the new two-stage auto-ignition prediction approach for all investigated fuels. Thus, it can be concluded that the approach proposed in this work also accounts for the influence of the fuel properties on the auto-ignition behavior in a correct manner.

It is important to remark that **neither the end-of-integration values of both ignition integrals representing the critical pre-reaction state and equaling one, nor the global auto-ignition reaction progress after the low-temperature ignition equaling 0.3 have been modified within the performed fuel and operating condition variations**. Hence, the reliable and accurate prediction of the time of mixture auto-ignition is ensured, even if proposed the two-stage approach is fully decoupled from the detailed mechanism and **no recalibration of the auto-ignition model constants is needed**. Consequently, any data available from calorific simulations, combustion analysis (PTA) or combustion simulations with a two-zone SI combustion model can be used directly for the prediction of auto-ignition in the unburnt mixture.

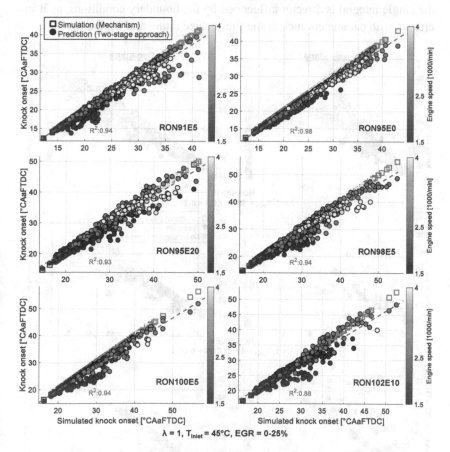

Figure 4.23: Crank angles of auto-ignition of single cycles simulated with a detailed mechanism and predicted by the developed two-stage approach for different fuels at various operating conditions.

Finally, Figure 4.24 demonstrates the huge gain in auto-ignition prediction accuracy achieved by accounting for the influence of the low-temperature heat release on the auto-ignition behavior of different fuel-air mixtures as well as the very limited prediction capabilities of the single stage auto-ignition prediction approach (Livengood-Wu integral) incorporated into all knock models commonly used today. Furthermore, it is obvious that the prediction error of the single integral is directly influenced by the boundary conditions, as it increases with parameters such as the fuel octane rating and engine speed.

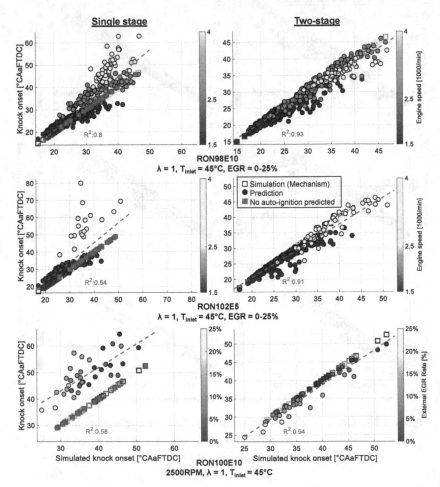

Figure 4.24: Auto-ignition prediction with the commonly used single stage (knock integral) and the newly developed two-stage approaches for two different fuels.

Overall, the performed extensive validation of the newly developed two-stage approach has demonstrated the high prediction capability of the model at various operating conditions. It can be concluded that the new auto-ignition prediction approach represents exactly the behavior of the detailed kinetic reaction mechanism at in-cylinder conditions.

4.5.4 Modeling the Influence of Injected Water on the Auto-Ignition Behavior

In order to meet the requirements posed by future SI engine concepts involving water injection, the modeling of the still missing influence of injected water on the auto-ignition behavior is presented in this section. One of the main benefits of water injection is that the high heat capacity of the injected water cools down the unburnt mixture, thus suppressing the occurrence of knock [48] [148]. The charge air and the combustion chamber are cooled by the water's vaporization enthalpy that is several times higher than the fuel's latent heat [72] [73] [81]. Thus, the knock propensity is reduced, resulting in an increased combustion efficiency in the high-load region. Alternatively, the cooling effect of injected water can be used for an increase of the engine compression ratio [72] [73]. Recent studies on this topic have further shown that the cooling potential of water strongly depends on the injection method [73] and, in case of direct injection, the injection timing [81]. Water further has the potential to minimize the need for high- and full-load enrichment [72] [73] [81] and decreases the pre-ignition tendency [72].

Figure 4.25 shows that the water content also directly affects the chemical processes leading to auto-ignition. Obviously, the influence of injected water on the three parameters characterizing the two-stage ignition phenomenon is not as pronounced as the effects of EGR or the AFR – an exhaust gas mass fraction of 5 %, an AFR of 1.3 and a water content equaling the total fuel mass have similar effects. Furthermore, such a high water content is expected to cause problems regarding the mixture flammability [48]. Nevertheless, it is obvious that in order to achieve maximum auto-ignition prediction accuracy in this case, the influence of injected water has to be accounted for, mainly because the addition of water leads to a considerable reduction of the temperature increase resulting from the low-temperature ignition that considerably influences the auto-ignition behavior of the mixture.

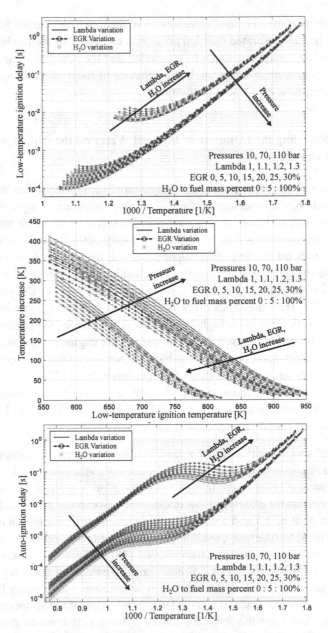

Figure 4.25: Influences of injected water, AFR and EGR on the three main characteristics of the two-stage ignition phenomenon.

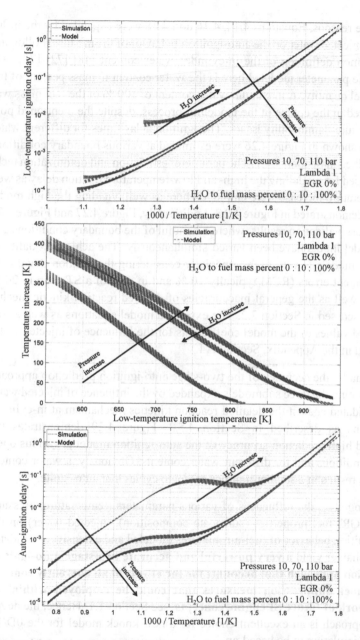

Figure 4.26: Simulated and modeled influence of injected water on the three main characteristics of the two-stage ignition phenomenon.

For these reasons, Equations 4.14, 4.16 and 4.18 were expanded by the influence of injected water on the auto-ignition behavior of the mixture. Following the common definition of the in-cylinder water content [48] [72] [73] [81] [148], the parameter modeled here is the water content in mass percent of the entire fuel quantity. A maximum water content of 100 % of the fuel mass was considered in the course of the modeling process, despite the mentioned possible mixture flammability issues. The ignition delay times for different water contents shown in Figure 4.26 were estimated at various boundary conditions as described in Section 4.4.1. The parameter estimation and optimization techniques used for modelling the high- and low-temperature ignition delay as well as temperature increase were employed here as well, yielding the high model quality demonstrated in Figure 4.26. Additionally, Figure 4.27 and Figure 4.28 show cross-validation results, where a handful of the boundary conditions and thus model inputs have been varied simultaneously. The achieved quality of all three models is very high, especially considering the ignition delay time measurement errors (RCM typically 10 % and in case of a ST – up to 20 % [24]) as well as the general inaccuracies of detailed reaction kinetic mechanisms discussed in Section 2.2. The expanded model equations as well as the estimated values of the model coefficients for the influence of injected water are listed in the Appendix, Section A4.

Additionally, the accuracy of the two-stage auto-ignition prediction approach incorporating the three submodels expanded by the influence of injected water was validated against the detailed reaction kinetics mechanism at in-cylinder conditions as described in Section 4.5.3. Figure 4.29 demonstrates the achieved high prediction accuracy of the auto-ignition model at various operating conditions and for different water contents. Obviously, a water content increase results in a smaller number of single cycles that auto-ignite.

Taken together, the influences of various parameters (temperature, pressure, AFR, EGR, fuel properties (surrogate composition), injected water) on the auto-ignition behavior of fuel-air mixtures modeled and extensively validated in this chapter yield a **very powerful and accurate two-stage auto-ignition prediction approach that accounts for the effects of all currently conceivable knock suppression measures that could be employed within the framework of future SI engine concepts**, see Section 1.1. Hence, the developed approach is an excellent choice for a new knock model for the 0D/1D engine simulation to be based on.

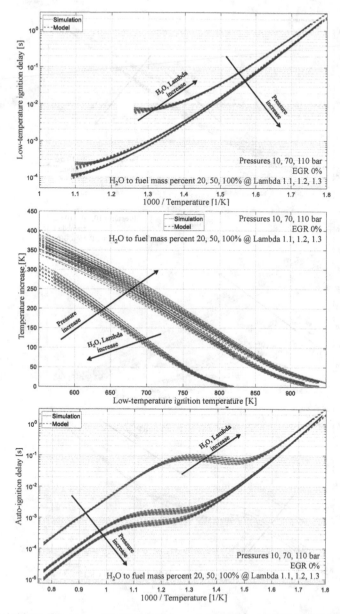

Figure 4.27: Cross-validation of the modeled influences of AFR and injected water on the three main characteristics of a two-stage ignition at different temperatures and pressures.

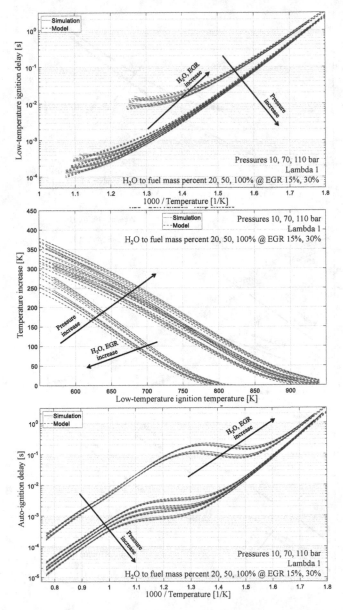

Figure 4.28: Cross-validation of the modeled influences of EGR and injected water on the three main characteristics of a two-stage ignition at different temperatures and pressures.

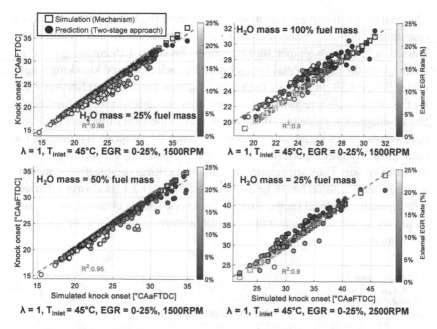

Figure 4.29: Validation of the modeled water influence on the auto-ignition behavior at in-cylinder conditions for the base fuel.

4.6 Two-Stage Approach Application to Measured Single Cycles

In the previous section, it has been demonstrated that the newly developed auto-ignition model can be used for the accurate prediction of local auto-ignition in the unburnt mixture. However, the permissibility of the two-stage approach underlying a new 0D/1D model for the prediction of the knock boundary and the knock onset in SI engines, provided that the exact values of all model inputs at the location in the unburnt mixture where knock is initiated ("knock-spot") are known, still has to be proven. To this end, investigations regarding the mixture inhomogeneities have been performed, as already done in Section 3.3.6 for the commonly used single stage knock integral. The main task here has again been simplified to the iterative estimation of the local temperatures at the knock-spots for each measured knocking single cycle.

The goal of the iterative calculation is to estimate the temperature offset representing a "hot-spot" needed so that the predicted time of auto-ignition expressed in °CA (high-temperature integral in Equation 4.19 equaling one) coincides with the experimental knock onset for each measured knocking single cycle, Figure 4.30. In contrast to the investigation performed in Section 3.3.6 for the single stage integral, here no measurement-series-individual reference end-of-integration value has to be defined, as in the case of the newly developed two-stage auto-ignition prediction approach the integration processes for both ignition stages is performed until the corresponding integral has reached the value of one. Thus, the requirement that the critical pre-reaction state representing auto-ignition has to be constant, Section 2.3.2.1, is always satisfied. Subsequently, the local temperature at the knock-spot is obtained as the sum of the mean unburnt temperature and the iteratively calculated temperature fluctuation.

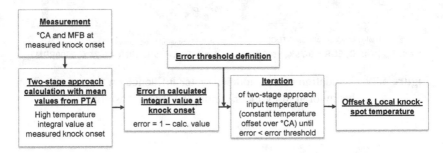

Figure 4.30: Iterative calculation of the temperature at the location where knock is initiated with the two-stage auto-ignition model.

To begin with, the sensitivity of the time of auto-ignition expressed in °CA to changes in temperature predicted by the two-stage approach was investigated at various operating conditions. Figure 4.31 shows the results for three single knocking cycles picked arbitrary for each investigated operating point. Obviously, exhaust gas reduces the approach's sensitivity to temperature changes because of the generally higher ignition delay times of both ignition stages as well as the smaller temperature increase after the first ignition stage it results in, Figure 4.11, Figure 4.13 and Figure 4.15. Because of the inherently higher unburnt temperature levels, increasing the engine speed generally reduces the approach's sensitivity to absolute temperature changes too (especially if the different limits of the x-axis in Figure 4.31 are considered). The presence of excess air reduces the sensitivity too; however, the decrease is not very significant because of the high temperature increase after the first ignition stage resulting from the generally lower unburnt temperature level caused by the slow combustion in this case.

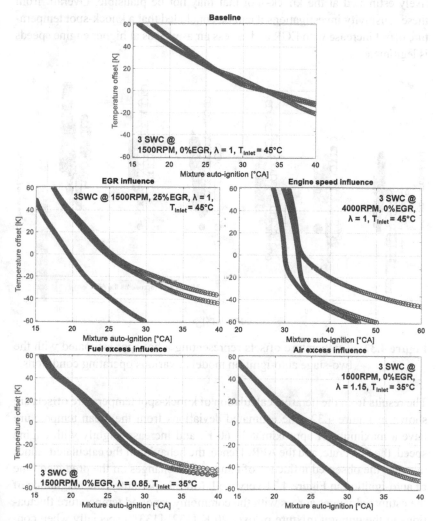

Figure 4.31: Influence of the operating conditions on the sensitivity of the two-stage auto-ignition prediction to changes in temperature.

Generally, a high temperature sensitivity of the two-stage approach means that
any calculation and / or input temperature errors can cause significant inaccuracies in the predicted time of auto-ignition. On the other hand, a low sensitivity to temperature changes could lead to too high temperature offsets iteratively estimated at the knock-spot that may not be plausible. Overall, from
these sensitivity investigations it can be concluded that a knock-spot temperature offset increase with EGR and excess air as well as at higher engine speeds
is legitimate.

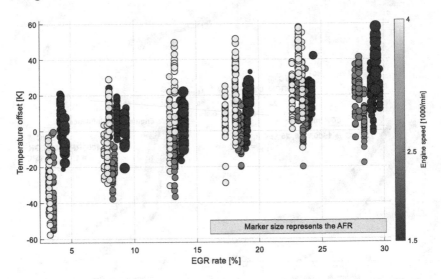

Figure 4.32: Temperature offsets representing a hot-spot estimated with the
two-stage auto-ignition model at various operating conditions.

The results from the iterative calculation of knock-spot temperature offsets are
shown in Figure 4.32. The estimated deviations from the mean temperature
have a maximum of approximately 60 K, and increase slightly with engine
speed, the EGR rate, and the AFR. Hence, the behavior of the calculated values
matches the observed influence of temperature changes on the predicted time
of auto-ignition in Figure 4.31 very well. More importantly, the magnitude of
the estimated values agrees with the commonly reported temperature fluctuations in the unburnt mixture of over 20 K [132] [133], especially when considering the fact that here single engine cycles have been investigated. Furthermore, the iteration results contain measurement and knock onset detection

inaccuracies, engine-specific effects, PTA model errors, as well as impreci-
sions of the two-stage approach and the sub-models for the three characteristic
parameters of the two-stage ignition. An additional inaccuracy results from the
assumption that the temperature offset does not change in time (constant value
over °CA). Moreover, in the course of the measurement data analysis it has
been assumed that the pressure pegs as well as the adapted cylinder mass values of the single cycles equal the peg of their corresponding averaged cycle,
Section 3.4, thus leading to a single cycle-individual temperature error at the
start of the PTA calculation. The observed negative offset values result from
the negative temperature coefficient behavior of gasoline fuels. They lead to a
change in the low-temperature ignition behavior and thus to higher temperature increase values that in turn cause a more rapid growth of the high-temperature integral value in Equation 4.19.

With these considerations in mind, it can be concluded that the estimated temperature fluctuation behavior is plausible and the magnitude of the calculated
values can be classified as realistic. It should be remarked that the corresponding calculations with the commonly used single stage integral in Section 3.3.6
yielded temperature offsets of partially well over 100 K, indicating that there
is a fundamental problem with the single stage approach, as such unburnt temperature fluctuations are not possible in a real engine.

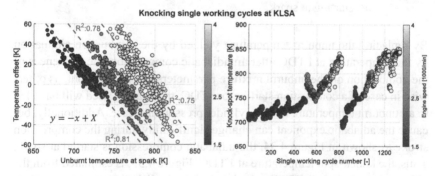

Figure 4.33: Estimated temperature offsets as a function of the unburnt temperature at spark and knock-spot temperatures at FTDC at various operating conditions.

As no quasi-dimensional models that can coupled with the Entrainment model
presented in Section 2.3.1.2 exist for the simulation of inhomogeneities, the

iteratively calculated temperature fluctuations have to be modeled empirically by fitting appropriate equations to the values estimated from the available measurement data. The calculated temperature offsets and unburnt temperature at spark correlate with a high goodness of fit, as shown in Figure 4.33. However, coupling the knock-spot and the unburnt temperatures at spark is not favorable for the simulation of spark timing sweeps. Therefore, based on this observation, the knock-spot temperature at FTDC shown in the right plot in Figure 4.33 can be related to the unburnt temperature at FTDC.

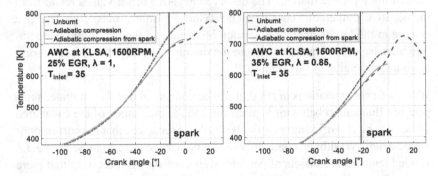

Figure 4.34: Unburnt temperature estimated with a two-zone combustion simulation, adiabatic compression and adiabatic compression starting at spark.

By replacing the unburnt temperature yielded by the combustion simulation by the temperature at FTDC after an adiabatic compression, the influences of the combustion on the unburnt mixture parameters can be ruled out, as otherwise in case of a combustion start before TDC, the heat released will have led to an unburnt temperature and in-cylinder pressure increase. Additionally, because the adiabatic exponent can change significantly during the compression stroke as shown in Figure 4.34, the adiabatic compression is started at spark. Thus, the knock-spot temperature at FTDC, Figure 4.33, is estimated from the unburnt temperature at FTDC calculated after an adiabatic compression starting at spark, yielding the temperature inhomogeneities model given by Equation 4.21.

$$T_{KS,FTDC} = T_{ub,spark} \left(\frac{V_{cyl,spark}}{V_{cyl,FTDC}} \right)^{\kappa_{spark}-1} + T_{off} \qquad \text{Eq. 4.21}$$

$T_{KS,FTDC}$ knock-spot temperature at firing top dead center [K]

$T_{ub,spark}$ unburnt temperature at spark [K]

$V_{cyl,spark}$ cylinder volume at spark [m^3]

$V_{cyl,FTDC}$ cylinder volume at firing top dead center [m^3]

κ_{spark} adiabatic exponent at spark [-]

T_{off} temperature offset representing the knock-spot [K]

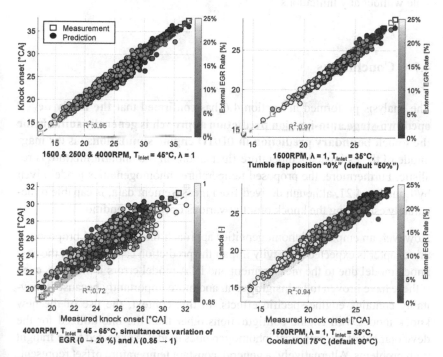

Figure 4.35: Measured knock onsets of single cycles and values predicted by the two-stage auto-ignition approach coupled with a knock spot model at different operating conditions.

Finally, the prediction performance of the proposed empirical temperature inhomogeneities model coupled with the newly developed two-stage auto-ignition prediction approach has been evaluated at various operating conditions, Figure 4.35. The model development and validation data appertain to two different datasets, as the single cycles used for the validation of the knock onset prediction have deliberately not been considered during the development of the inhomogeneities model. Obviously, the accuracy of the predicted knock onset values is very high, even when a handful of the operating conditions are changing simultaneously. This clearly proves that in case the exact values of all knock model inputs (boundary conditions at the knock-spot) are known at each integration step, the reliable prediction of knock occurrence based on both single and averaged working cycles with the two-stage approach is possible without any limitations.

4.7   Conclusions

The analysis performed in Section 4.6 has confirmed that **the newly developed two-stage auto-ignition prediction approach is generally suitable for the knock boundary prediction in 0D/1D engine simulations**, as the magnitude of the estimated temperature fluctuations in the unburnt mixture is realistic. Furthermore, the proposed temperature inhomogeneities model given by Equation 4.21, although derived from measurement data, is capable of accurately estimating the knock onset at various operating conditions.

However, an empirical inhomogeneities approach like the model proposed in this chapter is expected to heavily impair the prediction capabilities of the new knock model due to the measurement and PTA model errors it includes, even if these have proven to be insignificant, and more importantly because it inevitably contains engine-specific effects. In this context, the use of the new knock model with engine configurations other than the one utilized for the development of the empirical inhomogeneities model is expected to be fraught with problems. Alternatively, a general, constant temperature offset representing a hot-spot can be assumed, as studies suggest similar magnitude of the temperature fluctuation values. This constant offset can further be empirically modeled as a function of the operating conditions, e.g. engine speed. However, from the model developer's point of view, the simplest and currently most accurate strategy would be to predict the knock boundary whilst ignoring the temperature inhomogenieties, although studies have proven that these influence the knock behavior. Thus, neither empirical sub-models nor temperature fluctuation assumptions have to be included in the new knock model. More importantly, from the perspective of all currently available wall heat transfer approaches, mean unburnt temperature errors having the magnitude of the temperature fluctuations in the unburnt mixture (10 − 20 K) are considered negligible. Hence, the question arises, if an accurate prediction of the knock boundary is achievable with mean unburnt temperature values, which will be investigated in detail in Chapter 6.

Additionally, a local auto-ignition does not necessarily result in knock [88] [89]. For this reason, all investigations performed so far have been based on knocking single cycles. Hence, except the accurate prediction of the time of auto-ignition achieved by the development of the two-stage approach in this chapter, **a cycle-individual criterion for the occurrence of knock as a result from the auto-ignition is needed** and will be presented in Chapter 5.

5 Knock Occurrence Criterion

The transition to knock in SI engines does not only depend on the temperature and pressure in the end-gas [88]. The correct prediction of local auto-ignition is not sufficient for the reliable calculation of the knock boundary, as the occurrence of this phenomenon does not necessarily result in knock [88] [89]. To this end, an additional criterion for occurrence of knock resulting from the predicted auto-ignition is needed, which defines the main task in this chapter. Firstly, the possible modeling approaches for the knock probability decrease towards the end of combustion have to be reviewed.

5.1 Modeling the Reduced Knock Probability towards Combustion End

Except for not considering low-temperature ignition, commonly used knock models assume that no knock can occur after a pre-defined, constant MFB-point. In other words, the evaluation of the single knock integral is performed to an always-constant MFB-point, as it is assumed that auto-ignition after this point does not result in knock because of the small unburnt mass and volume fractions left, Section 2.3.2.2. Although it has been experimentally proven that the knock probability decreases towards the end of combustion, the selected fixed MFB-windows where no knock can occur are hardly linked to a physical process. They have been defined based on the evaluation of measurement data, which is also the reason why the values vary between the different knock models. Furthermore, the investigation of knocking single cycles measured at knock limited spark advance in Section 3.3.3 clearly shows that the heat release and thus MFB at the measured knock onset changes significantly with the operating conditions. Hence, a constant end-of-integration point, as assumed in the commonly used knock models, will lead to partially huge prediction errors. In Section 3.3.3, this has also been identified as one of the reasons for the poor prediction performance that 0D/1D knock models are known for.

As the latest MFB-point where knock can occur and thus the end-of integration point of all common knock models is constant, which results in poor prediction

© Springer Fachmedien Wiesbaden GmbH, part of Springer Nature 2019
A. Fandakov, *A Phenomenological Knock Model for the Development of Future Engine Concepts*, Wissenschaftliche Reihe Fahrzeugtechnik Universität Stuttgart, https://doi.org/10.1007/978-3-658-24875-8_5

performance, today's baseline for the most knock model users is a variable end-of-integration value $I_{k,crit}$ in the knock integral equation that represents auto-ignition, Section 2.3.2.2. By varying the critical pre-reaction state of the air-fuel mixture depending on the operating conditions (most commonly engine speed, sometimes cylinder mass at IVC), the prediction error resulting from the assumptions discussed in the previous paragraph can be compensated for. However, this yields a non-predictive knock model and disagrees with the main assumption of the Livengood-Wu integral in Equation 2.11 that at auto-ignition the integral value is always constant (equaling $I_{k,crit}$) and represents a critical concentration of chain carriers that causes the unburnt mixture left to auto-ignite. Hence, it is obvious that instead of varying the $I_{k,crit}$ level, a cycle-individual latest MFB-point where knock can occur has to be determined as a function of the operating conditions.

A straightforward solution to this task is to derive an empirical model for the end-of-integration MFB-point as a function of the operating conditions from measurement data. However, a simple adjustment of the approach used for the calculation of wall heat transfer, its parametrization or the cylinder wall temperature level will cause considerable changes in the MFB-values at the predicted time of auto-ignition as well as their behavior over the operating conditions, as will be demonstrated in Chapter 6. Furthermore, it is clear that an empirical sub-model derived from measurement data will contain engine-specific effects, measurement and pressure trace analysis model errors, Section 3.4. This will limit the prediction capabilities of the new knock model. Consequently, the flawless model use with different engine configurations is very unlikely. Hence, this empirical approach will not result in a considerable improvement of the knock prediction performance over the commonly used 0D/1D knock models.

Overall, it can be concluded that a measurement data fit of the end-of-integration point over the operating conditions is the simplest possible knock occurrence criterion; however, in sum it will heavily impair the prediction capabilities of the knock model. Clearly, only a purposeful knock occurrence criterion with a physical background that considers all operating conditions can improve the accuracy of the knock prediction. To this end, a phenomenological modeling approach considering the growing influence of the cold cylinder walls that is based on the unburnt mass fraction in the cool thermal boundary layer at the predicted time of auto-ignition will be presented in the next section.

5.2 Cycle-Individual Knock Occurrence Criterion

The engine cooling system leads to a temperature gradient in the gases close to the cool cylinder walls. This zone is commonly known as the thermal boundary layer and governs the cylinder heat transfer [8] [10] [75] [106]. The temperature of the gases in the boundary layer is much lower than the mean unburned mass temperature because of the cold walls. The flame reaches the cylinder walls at relatively late MFB-points, which coincides with the combustion phase where knock typically occurs in SI engines running on gasoline, Section 3.3.3. Hence, at the time of local auto-ignition in the unburnt mixture, a significant amount of the unburnt mass is within the thermal boundary layer and therefore cooler than the mean unburnt temperature, Figure 5.1. Thus, the thermal boundary layer has a significant influence on the knock behavior – a fact that can be used for the knock occurrence prediction.

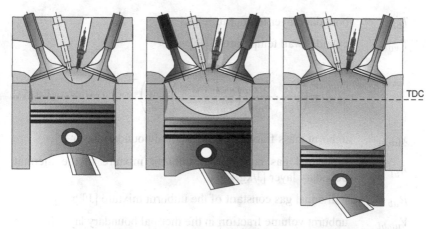

Figure 5.1: Exemplary boundary layer development on piston and liner over time and counteraction with the propagating flame.

After "entering" the boundary layer, the unburnt mixture cools down. This decreases the rate of the chemical reactions and the auto-ignition process is slowed down significantly. Based on this consideration, for the prediction of knock it can be assumed that if a considerable amount of the unburnt mixture is within the thermal boundary layer and thus has a temperature much lower than the mean unburnt temperature, the very small "hot" unburnt mass fraction left ($T = T_{ub}$) is unlikely to cause knock, even if it completely auto-ignites.

Hence, in the case of a significant unburnt mass amount within the thermal boundary layer, a local auto-ignition in the unburnt mixture cannot result in knock [142]. In the context of a knock criterion, this means that if the unburnt mass fraction in the boundary layer at the predicted time of auto-ignition is higher than a pre-defined threshold calibrated at the experimental knock boundary, no knock can occur.

Thus, the thermal boundary layer has to be modeled and the progress of its volume over time to be estimated. To this end, the boundary layer temperature is assumed to equal the mean value of the unburnt (estimated with a two-zone SI combustion model, Section 2.3.1.2) and the corresponding wall temperatures as shown in Equation 5.1 [106].

$$T_{bl} = 0.5 \cdot (T_{ub} + T_{wall}) \qquad \text{Eq. 5.1}$$

T_{bl} boundary layer temperature [K]

T_{wall} cylinder wall temperature [K]

$$x_{ub,bl} = \left(\frac{T_{bl}}{T_{ub}} \cdot \frac{R_{bl}}{R_{ub}} \cdot \left(\frac{1}{V_{ub,bl}} - 1 \right) + 1 \right)^{-1} \qquad \text{Eq. 5.2}$$

$x_{ub,bl}$ unburnt mass fraction in the thermal boundary layer [-]

R_{bl} individual gas constant of unburnt mixture in the thermal boundary layer [J/kg/K]

R_{ub} individual gas constant of the unburnt mixture [J/kg/K]

$V_{ub,bl}$ unburnt volume fraction in the thermal boundary layer [-]

$$x_{ub,bl} = \left(\frac{T_{bl}}{T_{ub}} \cdot \left(\frac{1}{V_{ub,bl}} - 1 \right) + 1 \right)^{-1} \qquad \text{Eq. 5.3}$$

The thermal boundary layer develops on liner and piston over time as a function of a handful of parameters as shown in Figure 5.1. The fraction of unburnt mass in the boundary layer at a specified point of time can be calculated with Equation 5.2, if the mean unburnt mixture temperature and the unburnt volume fraction in the boundary layer are known [142]. Generally, as the temperature within the boundary layer is lower than the unburnt temperature, the effective mass fraction in the boundary layer is higher than the volume fraction. Equation 5.2 can be further simplified by neglecting the ratio of the individual gas constants in the unburnt zone and the boundary layer, yielding Equation 5.3. The neglected ratio can be assumed to always equal one – an assumption that has been proven to result in a maximum error of less than 0.5 %

For the estimation of the unburnt volume fraction in the boundary layer, the combustion chamber is discretized in a finite number of small cylinders with a pre-defined height, Figure 5.2 left. Consequently, the boundary layer thickness is needed and can be estimated for a specific location on the cylinder wall at a given time with Equation 5.4, yielding the curves in Figure 5.2 right for different cylinder wall locations.

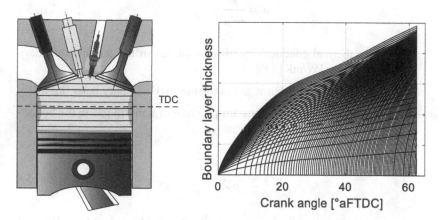

Figure 5.2: Combustion chamber discretization and development of the boundary layer at different cylinder wall locations over time.

Regarding the elapsed time in Equation 5.4, for locations below the top dead center, it begins at the moment they are uncovered by the piston, Figure 5.1 and Figure 5.2. Since there is uncertainty about the starting time for locations above the piston top-center position, this top-center position can be defined as the starting time for all locations in the clearance volume [106]. Hence, at these locations the boundary layer thickness and thus volume are always zero before firing top dead center.

$$\delta_t = 0.6 \cdot Re^{0.2} \cdot \sqrt{\alpha t} \qquad\qquad \text{Eq. 5.4}$$

δ_t boundary layer thickness at a specified cylinder wall location [m]

Re Reynolds number at the specified cylinder wall location [-]

α thermal diffusivity of the unburnt gas in the thermal boundary layer [m²/s]

t time elapsed since the beginning of the thermal boundary layer development at the specified cylinder wall location [s]

$$\alpha = \frac{k}{\rho c_p} \qquad\qquad \text{Eq. 5.5}$$

k thermal conductivity of unburnt gas in the thermal boundary layer [W/mK]

ρ density of unburnt gas in the thermal boundary layer [kg/m³]

c_p heat capacity at constant pressure of unburnt gas in the boundary layer [J/kg/K]

$$k = \frac{c_p \mu}{Pr} \qquad\qquad \text{Eq. 5.6}$$

μ viscosity of unburnt gas in the thermal boundary layer [Ns/m²]

Pr Prandtl number at the specified cylinder wall location [-]

The thermal diffusivity needed for the estimation of the thermal boundary layer thickness is calculated from the gas thermal conductivity, the density, and the heat capacity with Equation 5.5, where these gas properties are evaluated at the boundary layer temperature T_{bl}. However, investigations showed that the density as well as the heat capacities and their ratio γ of the unburnt mixture (and thus all evaluated at T_{ub}) can be used for the thermal diffusivity calculation too. This does not have a negative impact on the accuracy of the calculation results, as will be demonstrated in Chapter 6, and saves computational time, because all values of the unburnt mixture's parameters are calculated by the two-zone SI combustion model. Together with the neglect of the ratio of the individual gas constants yielding the simplified Equation 5.3 for the unburnt mass fraction in the boundary layer, this yields a knock occurrence criterion that involves no additional evaluations of any calorific gas properties at the boundary layer temperature. The influence of these model computational performance optimization measures on the knock boundary prediction accuracy will be discussed in detail in Chapter 6.

$$\mu_{air} = 3.3 \cdot 10^{-7} \cdot T_{bl} \qquad \text{Eq. 5.7}$$

μ_{air} viscosity of air fraction in the thermal boundary layer [Ns/m^2]

$$\mu_{EGR} = \mu_{air}/(1 + 0.027\phi) \qquad \text{Eq. 5.8}$$

μ_{EGR} viscosity of exhaust gas fraction in the thermal boundary layer [Ns/m^2]

ϕ fuel-air equivalence ratio[15] [-]

$$Pr = 0.05 + 4.2(\gamma - 1) - 6.7(\gamma - 1)^2, \phi \leq 1$$

$$Pr = \frac{[0.05 + 4.2(\gamma - 1) - 6.7(\gamma - 1)^2]}{[1 + 0.015 \cdot 10^{-6}(\phi T_{bl})^2]} \qquad \text{Eq. 5.9}$$

γ ratio of specific heats of unburnt gas in the thermal boundary layer [-]

The estimation of the thermal conductivity in Equation 5.6 requires the calculation of the gas viscosity and the Prandtl number. These are functions of the boundary layer temperature, the ratio of specific heats and the equivalence ratio ϕ and can be obtained with Equation 5.7, Equation 5.8 and Equation 5.9

---

[15] The fuel-air equivalence ratio ϕ is not to be mistaken with the air-fuel equivalence ratio λ, which is also known as relative air/fuel ratio. The two parameters are reciprocal [75].

[109]. An evaluation of the thermal conductivity with the correlations proposed in [121] as reported by [8] resulted in minor differences in the calculated absolute values, but did not show any considerable change of the parameter behavior over the operating conditions.

For the calculation of the last parameter needed for the estimation of the boundary layer thickness in Equation 5.4, which is the Reynolds number, Equation 5.10 is used. It estimates the value of the Reynolds number at a cylinder wall position with the coordinate x_0 (specific location on piston or liner). The gas velocity at the selected location x_0 results from a one-dimensional axial gas flow with velocity varying linearly from zero at the cylinder head to the piston velocity v_p at the piston crown, as shown in Equation 5.11 [106].

$$Re = \frac{\rho v x_o}{\mu} \qquad \text{Eq. 5.10}$$

v gas velocity at the specified cylinder wall location [m/s]

x_o coordinate of the specified cylinder wall (piston / liner) location [m]

$$v = v_p \left(\frac{x_o}{x}\right) \qquad \text{Eq. 5.11}$$

v_p piston velocity [m/s]

x current distance between the piston top and the cylinder head [m]

Thus, all parameters needed for the estimation of the boundary layer thickness in Equation 5.4 are known. The calculated results for one working cycle are shown in Figure 5.3. Generally, the highest layer thickness at each wall location x_0 is achieved at the end of the calculation because of the way the boundary layer develops with time. The assumed one-dimensional axial gas flow imposes that the fastest layer thickness growth is at wall locations near the current piston position. However, in this case there is less time available for the layer to develop, as the piston has just uncovered these locations. On the other hand, the assumed gas flow means that the greater the distance between piston and cylinder head, the slower the boundary layer development will be at locations above top dead center. These interrelations yield a trade-off regarding the layer thickness between time and location position x_0 and result in the highest boundary layer thickness values being achieved at locations near TDC.

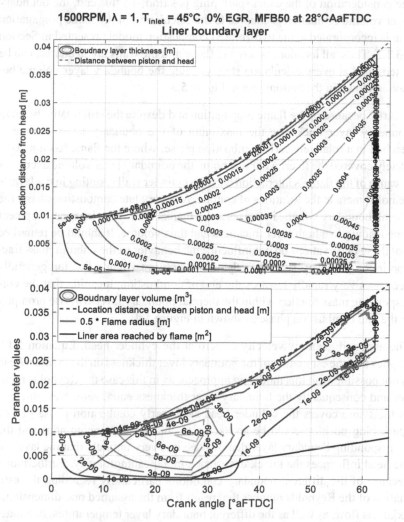

Figure 5.3: Estimated boundary layer thickness on the cylinder liner and influence of the propagating flame on the boundary layer volume for one engine cycle.

As soon as the propagating flame has reached the cylinder walls, it leads to a demolishment of the boundary layer at the locations of contact as shown in Figure 5.1. Hence, estimating the unburnt mass fraction in the boundary layer requires spatial information about the flame propagation over time, as well as the consideration of the exact spark plug position. To this end, the boundary layer volume estimation is coupled to the calculation of the flame propagation that is incorporated in the two-zone SI combustion model presented in Section 2.3.1.2. Thus, all locations where the flame and the walls are in contact can be determined at every calculation step, yielding the boundary layer volume behavior shown in the bottom plot in Figure 5.3.

Clearly, because of the flame propagation and despite the still relatively small boundary layer thickness, the maximum of the boundary layer volume is reached in a relatively early combustion phase, where the flame has not completely covered the liner. After this point, the boundary layer volume decreases because of the flame's interaction with the cylinder wall resulting in the layer's demolishment at the locations of contact. At a very late combustion phase, the overall boundary volume increases again thanks to the long layer development time leading to a high layer thickness and thus volume. It should be remarked that these observations do not influence the behavior of the unburnt mass fraction in the boundary layer directly, as the unburnt volume fraction generally decreases very rapidly towards the end of combustion, meaning that the corresponding mass fraction within the thermal boundary layer will rise promptly in the late combustion phase as shown in Figure 5.4.

The assumed gas flow velocity of zero at the cylinder head, Equation 5.11, implies that the estimation of the boundary layer thickness on the cylinder head is not possible with this modeling approach, as in this case the Reynolds number and consequently the boundary layer thickness equal zero. Nevertheless, as the flame covers the cylinder head at an early combustion phase (a side spark plug position could be an exception, but is very uncommon) and the corresponding boundary layer demolishes, it is very unlikely that the layer on the head influences the knock occurrence in a meaningful way. Furthermore, because of the different boundary conditions particularly regarding the estimation of the Reynolds number that result from the assumed one-dimensional axial gas flow, as well as the different boundary layer temperatures, two independent calculations for the piston and liner boundary layers are performed.

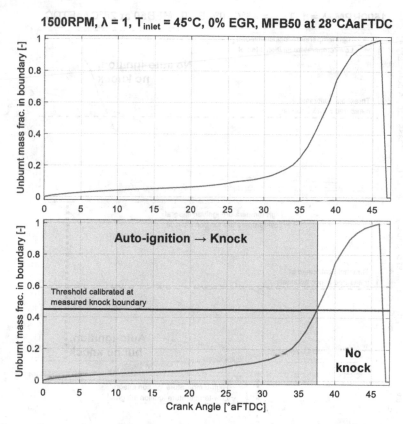

Figure 5.4: Exemplary unburnt mass fraction in the boundary layer curve and regions where an auto-ignition results in knock governed by the model calibration at the experimental knock boundary.

Generally, the thermal boundary layer development is calculated for every simulated cycle. At the same time, the independent prediction of local auto-ignition is performed with the newly developed two-stage approach and the corresponding sub-models presented in Sections 4.4 and 4.5. The layer estimation yields the progress of the unburnt mass fraction in the thermal boundary layer over °CA (or MFB) as shown in Figure 5.4. The rapid rise of the fraction in the late combustion phase primarily results from the fast decrease of the total unburnt volume.

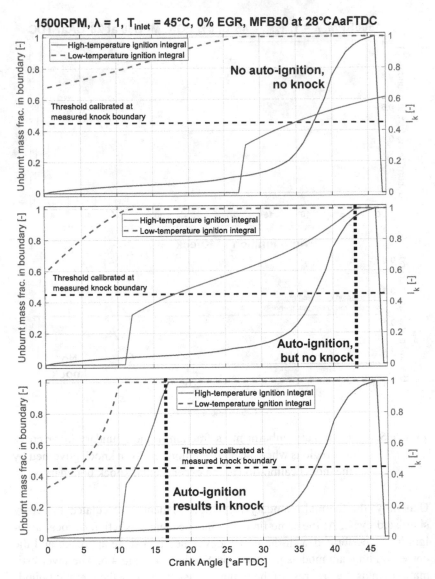

Figure 5.5: Prediction of knock occurring as a result from an auto-ignition
in the unburnt mixture with the proposed cycle-individual criterion based on the thermal boundary layer.

Subsequently, the unburnt mass fraction in the thermal boundary layer at the predicted time of local auto-ignition can be estimated. If the calculated value is bigger than a threshold calibrated at the experimental knock boundary, as shown in Figure 5.4 the occurred auto-ignition is assumed to not result in knock for reasons discussed at the beginning of this section. The pre-defined threshold for the unburnt mass fraction in the thermal boundary layer at the time of auto-ignition is the only calibration parameter of the new knock model. It is engine-specific and does not change with the operating conditions.

The proposed knock prediction strategy yields three possible cases regarding the interrelation between auto-ignition and knock occurrence that can be found in Figure 5.5:

- If the unburnt mixture is predicted to not auto-ignite, the boundary layer calculation is of no relevance for the knock prediction, as in this case no knock can occur.

- If the mixture auto-ignites in the late combustion phase, a significant amount of the unburnt mass will be within the cool thermal boundary layer at the time of auto-ignition. Thus, the occurred auto-ignition does not result in knock.

- Ultimately, if at the predicted time of auto-ignition the unburnt mass fraction in the boundary layer is smaller than the calibrated threshold representing the experimental knock boundary, the occurred auto-ignition results in knock.

Consequently, in case of a spark sweep starting at late combustion centers where no auto-ignition occurs in the unburnt mixture, the spark timing can be advanced, which will cause a gradual decrease of the unburnt mass fraction in the thermal boundary layer at the predicted time of auto-ignition, Figure 5.6. As soon as the estimated value falls below the calibrated threshold representing the experimental knock boundary, the corresponding spark timing results in knocking combustion and thus represents the knock limited spark advance. Further advancing the spark timing will result in auto-ignition in the early combustion phases (the extreme case is represented by an auto-ignition before combustion start, which corresponds to pre-ignition) and will yield even smaller unburnt mass fractions in the boundary layer at the time auto-ignition that are underneath the calibrated threshold corresponding to the knock boundary, implying the occurrence of severe knock. These interrelations can be used

for setting up a simple controller for the estimation of the knock boundary that
adjusts the spark timing of the combustion model.

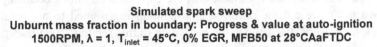

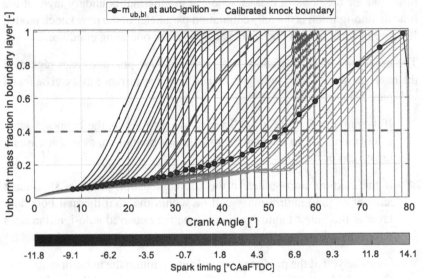

Figure 5.6: Spark timing sweep for the prediction of the knock boundary by
evaluating the unburnt mass fraction in the boundary layer at
auto-ignition.

5.3 Conclusions

The knock prediction approach proposed in this chapter **accounts for effects
of the current operating conditions**, as these influence both the boundary
layer development and the flame propagation significantly, **as well as for the
flame propagation** itself. Thus, the developed phenomenological knock occurrence criterion yields **a cycle-individual latest MFB-point where knock
can occur**. It contains **no empirical measurement data fits** and considers a
handful of cylinder geometry parameters influencing the boundary layer, such

as the top dead center clearance, bore, stroke, spark plug position, and the piston diameter. Thus, the new knock model **accounts for differences in the engine geometry** and has just **one calibration parameter** that is engine-specific and does not change with the operating conditions. Therefore, after a simple recalibration, the new, **fully predictive knock model** can be applied to different engines without any limitations, as will be demonstrated in Chapter 7.

6 Knock Model Overview

The present chapter gives an overview of the developed 0D/1D knock prediction approach as well as some remarks on the model use. To begin with, the
possible sources for the model inputs needed for the estimation of the knock
boundary are reviewed and the prediction quality they result in is evaluated.
Additionally, the model versatility is investigated to ensure that the developed
approach can be used in combination with various simulation models for the
estimation of parameters such as the wall heat transfer. Finally, the model
workflow in the course of the knock boundary prediction, the model outputs,
and calibration process are described in detail.

6.1 Discussion on the Knock Model Inputs

Commonly, the mean values of the unburnt zone's parameters estimated with
an appropriate two-zone SI combustion modeling approach, in this work the
Entrainment model as described in Section 2.3.1.2, are used as knock model
inputs. The main problem when simulating knock within a 0D/1D environment is the fact that the knock model calculations are performed with values
calculated by other models. Hence, the model errors of all other models are
reflected in the inputs of the knock model. This requires the very careful calibration of all simulation models involved. On the other hand, in order to guarantee the model usability, the full functionality should to be ensured even if
the values of the knock model inputs are inaccurate because of flaws of other
simulation models or errors in their parametrization. As an example, in case of
badly estimated cylinder wall temperatures or poorly parametrized wall heat
transfer approach, the unburnt temperature level will be imprecise. To this end,
the following sections address some use case scenarios demanding high model
flexibility as well as their influence on the knock boundary prediction accuracy.

Performed investigations showed that generally, the higher the unburnt temperature level is, the earlier auto-ignition will occur, and therefore the smaller
the unburnt mass fraction in the boundary layer will be at auto-ignition. Thus,

© Springer Fachmedien Wiesbaden GmbH, part of Springer Nature 2019
A. Fandakov, *A Phenomenological Knock Model for the Development of
Future Engine Concepts*, Wissenschaftliche Reihe Fahrzeugtechnik Universität
Stuttgart, https://doi.org/10.1007/978-3-658-24875-8_6

inaccuracies in the unburnt temperature level (no matter the reason) are reflected in the knock model calibration parameter $x_{ub,bl}$ and can be compensated for by adjusting its value in the course of the knock model calibration. In certain cases, this, together with the fact that the knock simulation is always performed based on the average cycle as will be discussed in Section 6.1.5, can result in perhaps unrealistic MFB-points at auto-ignition as well as thresholds for the percentage of unburnt mass in the boundary layer of about 10 – 15 % at the experimental knock boundary. Hence, the absolute values of the calibration parameter $x_{ub,bl}$ strongly depend on the parametrization and calculation accuracy of many other simulation models, as will be demonstrated in the following sections. Thus, the estimated absolute values of the unburnt mass fraction in the thermal boundary layer as well as the corresponding times of auto-ignition (expressed in °CA or MFB) are not always meaningful. Consequently, these should not be interpreted, although the knock model calibration parameter $x_{ub,bl}$ has a physical background, as it describes a mass fraction of unburned fuel in the cool thermal boundary that is high enough, so that the probability of auto-ignition of unburned fuel not in the boundary becomes small as well.

6.1.1   Consideration of Inhomogeneities

Generally, 0D/1D simulations do not account for pressure fluctuations, meaning that there is no difference between the pressures in the burnt and unburnt zones in case of a two-zone combustion model, Section 2.3.1.2. The temperature is by far the most critical knock model input, as it has an exponential influence on the ignition delay times of both ignition stages, as has been discussed in Sections 2.3.2.1, 4.4 and 4.5. Thus, small differences in the temperature values can cause significant changes in the auto-ignition behavior predicted by the newly developed two-stage approach given by Equation 4.19. In this context, the question arises, if and how the mixture and temperature inhomogeneities in the unburnt mass that influence the knock behavior and are generally known as "hot-spots", Section 3.3.6, should be accounted for, as already done by applying the two-stage auto-ignition prediction approach to measured single cycles in Section 4.6. Such temperature fluctuations with amplitudes of well above 20 K can occur even in a nominally homogeneous engine [132] [133]. However, currently no quasi-dimensional models that can coupled with the Entrainment model presented in Section 2.3.1.2 exist for the simulation of inhomogeneities in the unburnt mixture. Hence, there are two possible options for considering the temperature fluctuations:

■ Derivation of an empirical inhomogeneities model from measurement data, most appropriately knocking single cycles, yielding the model proposed in Section 4.6. As is the case with the knock occurrence criterion presented in Chapter 5 and despite the very high knock onset prediction accuracy achieved with the proposed empirical inhomogeneities model, this approach is expected to heavily impair the prediction capabilities of the knock model mainly because of engine-specific effects, but also due to the measurement and PTA model errors it contains. In this context, the use of the knock model with engine configurations other than the one utilized for the development of the empirical inhomogeneities model is expected to be fraught with problems, as significant inaccuracies resulting from the incorporated empirical temperature fluctuations approach might occur. For these reasons, this option shall not be considered for the new, fully predictive 0D/1D knock model.

■ The assumption of a general temperature offset representing a hot-spot, as experimental studies [132] [133] [135] as well as 3D CFD stratification analysis [28] suggest similar magnitude of the fluctuation values usually in the region of 10 to 20 K. This constant offset can further be empirically modeled as a function of the operating conditions, e.g. engine speed.

However, from the model developer's point of view, the simplest and currently most accurate strategy would be to predict the knock boundary whilst ignoring the inhomogenieties, although studies have proven that these influence the knock behavior. Thus, neither empirical sub-models nor fluctuation assumptions have to be included in the new knock model. Furthermore, the mean unburnt temperature is calculated with phenomenological models and thus, it reflects changes in the operating conditions, combustion system or engine configuration in a correct way, which is not necessarily the case with the inhomogenieties model derived from measured single cycles in Section 4.6 and also does not apply to an assumed constant temperature fluctuation value anyway. More importantly, from the perspective of all currently available wall heat transfer approaches, e.g. [8] [10] [169] mean unburnt temperature errors having the magnitude of the temperature fluctuations in the unburnt mixture (10 – 20 K) are considered negligible. Hence, the estimated unburnt temperature level will never coincide with the real one throughout different operating condition variations with such an accuracy that the values of the temperature fluctuations become relevant. For these reasons, the explicit consideration of temperature fluctuations in the course of a knock boundary prediction is not necessarily expedient.

Hence, the question arises, if an accurate prediction of the knock boundary is achievable with mean unburnt temperature values. To this end, Figure 6.1 compares the prediction errors resulting from simulations with the mean unburnt temperature and an assumed constant temperature offset of 15 K representing a hot-spot. Clearly, because of the different simulation types (with and without consideration of temperature fluctuations), prior to performing predictive simulations, the knock model has to be calibrated individually for each of the two investigated cases at the same measured knock boundary. This results from the fact that accounting for inhomogeneities influences the inputs of the knock model, as the temperature at the considered hot-spot is higher than the mean unburnt temperature. Thus, at the same experimental knock boundary, auto-ignition is estimated to occur earlier and the unburnt mass fraction in the thermal boundary layer at the time of auto-ignition slightly decreases if temperature fluctuations are accounted for, Figure 6.1.

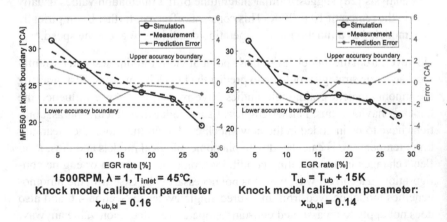

Figure 6.1: Knock boundary prediction with mean unburnt temperature values and an assumed constant temperature offset of 15 K representing a hot-spot.

Obviously, the fluctuation of the knock prediction error increases slightly, if the inhomogeneities are accounted for with a constant temperature offset. This effect is reasonable, as in reality, the assumed general value of 15 K is surely not constant, but rather cycle-individual. More importantly, Figure 6.1 clearly shows that the consideration of temperature fluctuations does not result in a higher prediction accuracy compared to the knock boundary simulation based on the mean unburnt temperature. Numerous simulations of further operating points and measurement series have shown similar results. Hence, thanks to the cycle-individual knock occurrence criterion proposed in the Chapter 5, the prediction errors achieved with mean unburnt temperature values are generally small enough for the accurate knock boundary simulation, as will also be demonstrated in the course of the knock model validation in Chapter 7. Furthermore, as already mentioned, the mean unburnt mass temperature, being estimated with phenomenological simulation models, always reflects changes in the operating conditions, combustion system and engine configuration.

Therefore, the mean unburnt temperature can be classified as the currently most reliable and thus suitable knock model input. For similar reasons, the knock boundary prediction is performed with the global AFR and exhaust mass fraction. Thus, the new knock model does not include any empirical sub-models. However, the high model sensitivity to temperature changes also has to be considered in the context of the estimation and parametrization of the cylinder wall temperatures as well as the choice of wall heat transfer approach for the simulation.

6.1.2 Choice of Wall Heat Transfer Approach

The wall heat transfer approach has a significant influence on the unburnt temperature level and thus on the auto-ignition prediction, as well as on the boundary layer temperature and volume, Section 5.2. Hence, it is obvious that the value of the knock model calibration parameter $x_{ub,bl}$ can vary significantly depending on the wall heat transfer approach selected for the simulation. Nevertheless, Figure 6.2 demonstrates that similar quality of the knock boundary prediction can be achieved with all three investigated wall heat transfer approaches commonly used today [8] [10] [169]. Thus, it can be concluded that **the choice of wall heat transfer approach does not influence the achievable knock boundary prediction quality**.

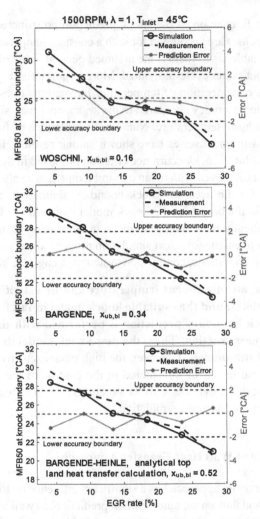

Figure 6.2: Achievable accuracy of the knock boundary prediction with different wall heat transfer approaches.

6.1.3 Cylinder Wall Temperatures and Unburnt Temperature Level

Except for the approach used for the calculation of wall heat losses, the cylinder wall temperatures also influence the unburnt temperature level significantly. For a reliable knock boundary prediction, it is essential that all changes of the wall temperatures with the operating conditions (e.g. increase with engine speed) are considered correctly.

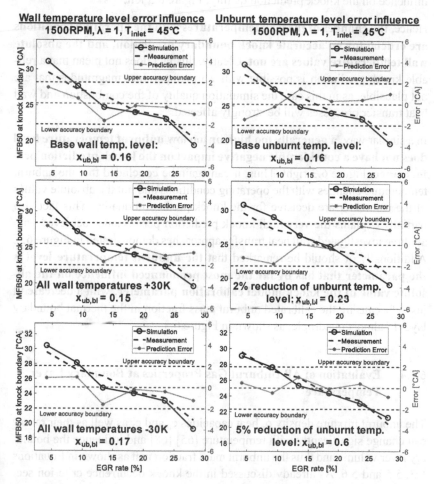

Figure 6.3: Achievable accuracy of the knock boundary prediction with different wall temperature parametrizations and at various unburnt temperature levels.

However, if the wall temperature level is generally too high or too low (e.g. because of bad parameterization of an imposed temperature or the wall thermal properties), this fact results in a change of the knock model calibration parameter $x_{ub,bl}$ estimated at the experimental knock boundary, but has no significant influence on the knock prediction quality, Figure 6.3 left.

Hence, **the changes of the wall temperatures with the operating conditions are crucial for an accurate knock boundary prediction, and the absolute wall temperature values are not.** Clearly, this fact does not mean that a reliable knock prediction is possible with wall temperature magnitudes that are not plausible, as in this case the simulation quality of the combustion and wall heat transfer processes will be strongly affected.

In a similar way, **a generally too high or too low unburnt temperature level does not have a considerable negative impact on the knock prediction performance**, Figure 6.3 right. Thus, it can again be concluded that the unburnt temperature changes with the operating conditions and not the absolute values of the parameter are decisive for the knock prediction quality. This fact also partially explains the very high knock prediction performance achieved with all three investigated wall heat transfer approaches demonstrated in Figure 6.2. Additionally, it should be remarked that **the unburnt temperature level is the parameter that has by far the most pronounced influence on the absolute value of the knock model calibration parameter**, with a decrease of just 5 % causing the calibrated unburnt mass fraction in the thermal boundary layer to almost quadruple, as shown in Figure 6.3.

6.1.4   Evaluation of All Unburnt Gas Properties at Boundary Layer Temperature

The unburnt mixture density ρ, heat capacities c_p and c_v as well as their ratio γ can change significantly with temperature [65] [68] and influence the boundary layer volume and thus the unburnt mass fraction in it as shown in Equations 5.3, 5.5 and 5.6. As already discussed in the knock occurrence criterion section, all unburnt gas properties needed for the boundary layer development calculation are supposed to be evaluated at the layer temperature T_{bl} [106]. However, the additional calculations of the density and heat capacities at T_{bl} result in an increase in computational time, as discussed in Section 5.2. In this context, Figure 6.4 shows that the calorific properties evaluation temperature does not affect the knock prediction quality considerably, as its effect can be compensated for by adjusting the knock model calibration parameter value in the course of the model calibration process. This is because an estimation of ρ, c_p, c_v and γ at the boundary layer temperature yields results in slightly different absolute values of the parameters, but their progress over °CA and MFB respectively is not significantly influenced by the evaluation temperature. This results from the fact that the mean unburnt temperature is also decisive for T_{bl} as given by Equation 5.1, so that the two parameters behave in a similar way.

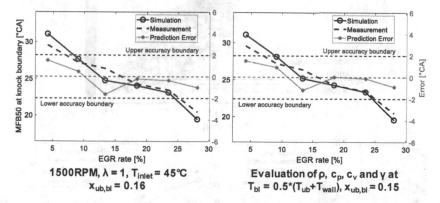

Figure 6.4: Influence of the calorific value evaluation temperature on the knock boundary prediction.

Hence, in order to save valuable computational time and thus optimize the computational performance of the model, **the knock boundary prediction can be performed with the density and heat capacities of the unburnt mixture yielded by the combustion model calculations, without negatively influencing the knock prediction quality**.

6.1.5 Simulation of Single Cycles

Generally, averaging a signal leads to an information loss [161]. Hence, not all data present in single engine working cycles is contained in the corresponding representative average cycle. This is particularly important for the knock boundary prediction because of the high knock model sensitivity to temperature changes. For this reason, a reproduction of the cyclic combustion fluctuations and the associated pressure and temperature curves seems to be a promising approach for improving the quality of the knock boundary prediction.

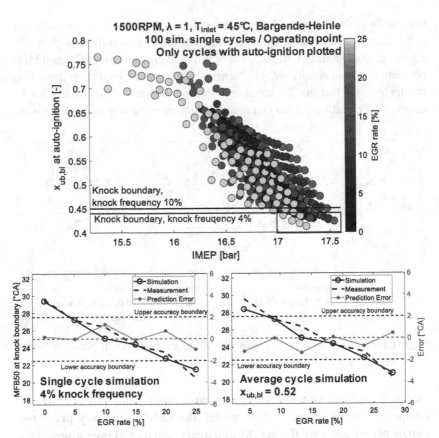

Figure 6.5: Achievable accuracy of knock boundary prediction based on simulated single and average working cycles.

To this end, investigations with a cycle-to-cycle variations simulation model presented in Section 2.3.1.3 were performed. The corresponding results are shown in Figure 6.5. Obviously, the knocking single cycles can indeed be identified (marked by a rectangle in Figure 6.5). Moreover, in the case of single-cycle simulation, the knock frequency (window) defining the knock boundary can be used directly as a knock model input. However, it is obvious that no significant gain in the prediction quality is achieved, presumably because of the still missing information about the mixture inhomogeneities and the temperature fluctuations in the unburnt mixture of the individual cycles that have to be estimated with an empirical model as proposed in Section 4.6.

Nevertheless, for reasons already discussed in detail in Section 6.1.1, the newly developed model performs the knock boundary prediction based on mean unburnt temperature values and any empirical modeling of unburnt temperature fluctuations is precluded. Furthermore, in case of a single-cycle-based knock prediction, **the achieved accuracy disadvantageously depends on the performance of one more model (CCV) and the quality of its parametrization.**

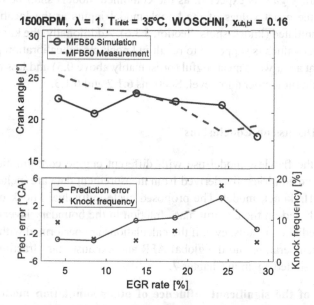

Figure 6.6: Relationship between the measured knock frequency and the knock boundary simulation error.

In average-cycle-based knock boundary simulations, minor changes in the measured knock frequency result in slightly different knock model calibration parameter values. Thus, the prediction error and the knock frequency show similar behavior, if the value of the calibration parameter $x_{ub,bl}$ is kept constant, Figure 6.6. Hence, average cycle simulations also indirectly account for the knock frequency (window) defining the knock boundary. This fact has to be considered during the knock model prediction performance assessment in Chapter 7, as the measured KLSA has been defined as a knock frequency window in Section 3.1.

Overall, the performed investigations show that the **single-cycle simulation is currently not beneficial for the knock boundary prediction and leads to a huge increase in computational time**. For this reason, the knock simulation performed with the new model is always based on the representative average cycle, even if an approach for the estimation of the cyclic combustion fluctuations is active. However, by coupling a phenomenological inhomogeneities model with a CCV simulation approach in the future, a further knock simulation accuracy gain is expected, as the combined models shall be capable of yielding the exact values of the knock model inputs at the location where knock is initiated (knock-spots, Section 3.3.6). Additionally, the knowledge of these exact values is supposed to result in knock model calibration parameter values that are always meaningful (presumably above 0.5) and less dependent on the unburnt temperature level, Sections 6.1.2 and 6.1.3.

6.1.6   Discussion Conclusions

Overall, the flawless model use with different engine configurations is ensured, as no empirical fits derived from measurement data are included in the new 0D/1D knock model. The proposed cycle-individual knock occurrence criterion based on the unburnt mass fraction in the boundary layer yields accurate prediction results, even if the calculations are performed with the mean unburnt temperature and the global AFR and exhaust mass fractions, as will be demonstrated again in Chapter 7.

Because of the significant influence of other simulation models on the knock model calibration parameter, its absolute value is not meaningful and should not be interpreted, despite its physical background. Overall, the unburnt temperature level is the parameter with by far the most pronounced influence on value of the calibrated unburnt mass fraction in the thermal boundary layer at the time of auto-ignition. Nevertheless, the performed investigations have shown that a similar prediction quality is achievable with different wall heat transfer approaches and cylinder wall temperature parametrizations, although these influence the unburnt temperature level significantly. The simulation of single cycles currently does not improve the knock boundary prediction accuracy and leads to a significant increase in computational time.

In general, **the most important aspect of simulating the knock boundary is consistency**. This means that if models yielding parameters that are inputs of the knock model are recalibrated or changed, the knock model has to be recalibrated as well. Of course, this is not to be mistaken with performing operating condition variations or changing the engine configuration. A simple example: in case of a wall heat transfer approach change in the course of engine operating point simulations, the knock model has to be recalibrated at one operating point at the experimental knock boundary, Figure 6.2. However, these reflections are true for many other simulation models as well, e.g. the two-zone SI combustion model.

6.2 Surrogate Composition Estimation

Since real gasoline fuels are composed of a large variety of hydrocarbon components, surrogate mixtures of only few representative species are typically employed in computational research and in engine development [54]. Furthermore, only one, specific surrogate composition is representative for a given gasoline fuel and matches up to its characteristics. As discussed in Section 4.2, the gasoline surrogates employed in the course of the knock model development are composed of n-heptane, iso-octane, toluene, and ethanol.

Unlike the knock models commonly used today which typically account for the RON of the real gasoline fuel, Section 2.3.2.2, the approach proposed in this work considers the effects of the surrogate component fractions on the three main characteristics of the two-stage ignition phenomenon, Sections 4.4.3, 4.4.4 and 4.4.5. The reason for this are the recent findings suggesting that having regard to just one real fuel property results in poor representation accuracy of the fuel's influence on knock occurrence, Sections 2.2 and 4.2.

Hence, at the beginning of the simulation, the respective representative surrogate composition corresponding to the real fuel selected for the knock boundary prediction has to be calculated. Thus, the influence of the estimated component fractions on the ignition delay of both ignition stages as well as on the temperature increase resulting from the low-temperature ignition can be considered. In this respect, the values of the fuel RON, MON, H/C ratio, and liquid density are all of interest due to their great importance for the accurate description of the real gasoline's auto-ignition behavior. Hence, a surrogate that re-produces equally well as many characteristics of the real fuel as possible is expected to be a much more accurate representation than a composition that perfectly matches only a couple of specified real gasoline properties. This re-quires the numerical optimization of the estimated composition by minimizing all property differences between the real and surrogate fuels [54]. On the other hand, an available detailed fuel analysis containing the exact values of all real fuel characteristics is rarely the case. Commonly, only a handful of these are available to the simulation engineer.

For these reasons, an iterative surrogate composition calculation was devel-oped based on various blending rules and subsequently integrated into the new knock model as an additional submodel that estimates the surrogate composi-tion prior to the knock boundary prediction from arbitrary real fuel character-istics specified by the user. Possible inputs are the fuel RON, MON, octane sensitivity, liquid density, H/C ratio, lower heating value (LHV), ethanol con-tent in volume percent, C/H/O atoms per molecule as well as the C/H/O mass fractions. The high number of possible user inputs significantly simplifies the definition of the fuel for the knock boundary prediction as well as the investi-gation of the influence of fuel properties on the knock behavior. Thus, the gen-erally limited availability of exact real fuel property values does not thwart the consideration of the gasoline characteristics in the course of the knock bound-ary prediction. Of course, the high degree of freedom provided to the knock model user poses some challenges to the development of such a surrogate es-timation submodel. The implemented iterative calculation accounts for the fact that it is important the surrogate reproduces equally well as many properties of the real fuel as possible due to their relevance to the auto-ignition behavior.

The developed approach incorporates various blending rules – linear and non-linear, mole-, volume- and mass-fraction-based. As the fractions of four sur-rogate components have to be estimated, an equation system consisting of four equations has to be formulated and solved iteratively. The first equation is given by setting the sum of all four fractions to one. Additionally, the ethanol content in volume percent can be constituted as a required user input, as this property value is commonly known (or can be guessed quite accurately), yield-ing the second system equation. Alternatively, the ethanol fraction can be es-timated from known O mass fraction or number of O atoms per molecule. If similar information is available for the H and C molecules too, the H/C ratio of the gasoline fuel can be calculated as well.

The two system equations left require two arbitrary real fuel characteristics to be specified by the user. These are then employed for the surrogate component fraction estimation with the respective blending rules:

■ The advanced, non-linear equations for the RON and MON discussed in Section 4.2 were inverted and analytically solved, so that (a) relationship(s) between the four surrogate components can be derived from the real fuel's RON and / or MON. Additionally, the relationship between the octane numbers and the octane sensitivity was implemented as well.

■ For the estimation of the liquid density, the components are blended linearly based on their volume fractions.

■ In case the H/C ratio of the real fuel is known, a linear mole-fraction-based equation is used.

■ The lower heating value of the surrogate is estimated with a linear mass-fraction-based blending rule.

The values of the liquid densities, lower heating values, H/C ratios and the molar masses of the four surrogate components are of common knowledge and can be found in various publications.

Because of the incorporated non-linear equations, the formulated equation system has multiple solutions. Clearly, as a first step, all negative solutions as well as those higher than one can be discarded because a surrogate component fraction has to be between zero and one. For the case that the iterative calculation yields more than one plausible solutions of the equation system (all four fractions are between zero and one), an additional plausibility check was implemented. It is based on a comparison of the surrogate and real fuel RONs, as in the common case, the RON of the real gasoline fuel is known or can be guessed quite accurately.

Thus, the iterative calculation of the surrogate composition requires four real fuel properties to be available (or (partially) guessed) in total, with two of them being the RON and the ethanol content in volume percent. The uncomplicated fulfillment of this requirement is realistic, especially because some real fuel property values are needed anyway for the estimation of the fuel mass entering the engine cylinder(s) in a 0D/1D simulation environment. Alternatively, the knock boundary simulation can be performed with the default surrogate composition proposed in Section 4.2.

Additionally, a prioritization algorithm was incorporated into the submodel. Based on the submodel inputs entered by the user, it determines which of the available real fuel properties have to be reproduced more accurately by the surrogate than others. The algorithm is activated, if more than four real fuel characteristics are available. Thus, with its high number of possible user inputs, the implemented submodel significantly simplifies the definition of the fuel for the knock boundary prediction as well as the investigation of the influence of fuel properties on the knock behavior.

6.3 Knock Model Use

After its development and implementation, the code of the new knock model was converted from MATLAB [110] to FORTRAN and consequently integrated into the FVV Cylinder Module as a post-processing module (without the influence of injected water, Section 4.5.4, and the surrogate composition calculation from real fuel properties in Section 6.2) [66]. Additionally, it was coupled with the flame propagation calculation incorporated into the Entrainment model, so that the contact points of the flame with the cylinder walls can be estimated as described in Chapter 5. As the impacts of the low-temperature heat release are only considered during the course of the knock boundary prediction within the new knock model, none of the models already available in the FVV Cylinder Module had to be modified. Furthermore, an interface of the FKFS UserCylinder [125] was used to perform simulations with the new knock model in combination with the commercial software GT-POWER [69]. Finally, the knock model code was tuned for increasing the computational speed.

This section gives an overview of the optional user inputs of the knock model as well as its outputs and discusses the calibration procedure. Additionally, the complete model workflow is presented. Finally, some post-processing techniques aiming at improving the model's computational and prediction performance are presented.

6.3.1 User Inputs and Model Calibration

The newly developed knock model has just one calibration parameter. Thus, it requires one user input – the unburnt mass fraction in the thermal boundary layer at the time of auto-ignition at the experimental knock boundary. Hence, the knock model calibration procedure has to be performed for at least one operating point at the experimentally estimated knock boundary.

One of the main challenges when simulating knock in a 0D/1D environment is the fact that the knock boundary prediction is performed based on parameter values calculated by other models. The calibration process starts with the careful adjustment of all other models, e.g. two-zone combustion, wall heat transfer, turbulence etc., so that the simulation accurately reproduces the measured cylinder pressure as well as the unburnt temperature and heat release rate from the combustion analysis (PTA or TPA, Section 2.3.1). This step is particularly important because of the fact that any errors in the calibration of all these models are reflected in the inputs of the knock model.

Subsequently, the new model can be activated for the simulation of one or several operating points at the experimental knock boundary that have been selected for the model calibration. During the course of this simulation, the newly developed knock prediction approach will evaluate if two-stage ignition occurs and will calculate the time of mixture auto-ignition as well as the progress of the boundary layer volume. Thus, the unburnt mass fraction in the thermal boundary layer at the time of auto-ignition that corresponds to the experimental knock boundary is obtained. Finally, this estimated value has to be entered as a knock model user input for the predictive simulation of various combustion systems, engine configurations, and / or operating conditions. **Overall, the adjustment steps required prior to performing predictive simulations with the new knock model do not differ in any way from the process of calibrating any commonly used knock model that is founded on the knock integral.** Unlike all other models however, the calibration parameter of the approach developed in this work describes a physical quantity.

By default, the knock prediction is performed with the surrogate composition corresponding to the RON96.5E10 gasoline selected as base fuel as described in Section 4.2. However, the exact composition of a specific surrogate representing another gasoline fuel and estimated with appropriate blending rules can be entered as an optional knock model user input. Thus, the influence of the used fuel and its properties on the knock occurrence can be considered in the course of the knock boundary prediction.

A further optional knock model user input is the activation of the evaluation of all unburnt mixture properties at the temperature of the boundary layer as discussed in Section 6.1.4, despite the recommended knock boundary prediction based on the density and heat capacities of the unburnt mixture yielded by the combustion model calculations. Furthermore, although not recommended, the consideration of the exact spark plug position for the calculation of the locations of contact between the propagating flame and the cylinder walls can be turned off. Additionally, scaling factors for the end values of the two ignition integrals (default value 1) as well as custom the possibility to change the starting value of the high-temperature ignition integral (default value 0.3 as discussed in Section 4.5.1) have been implemented. However, it is strongly recommended not to change these values. Moreover, the unburnt temperature level can be adjusted, in case it is generally too low or too high, in order to obtain realistic values for the time of auto ignition in the unburnt mixture (expressed in MFB) and the knock model calibration parameter, see Section 6.1.3. Finally, the implemented post-processing of the unburnt mass fraction in the thermal boundary layer (curve normalization and smoothing, both discussed in detail in Section 6.3.3) can be modified or completely deactivated.

Based on numerous knock prediction performance investigations, a **default dataset for all optional user inputs** has been defined:

- Surrogate composition in mass fractions representing the RON96.5E10 fuel used for the experimental investigations on the engine test bench: 0.457 iso-octane (*iO_m_frac*), 0.136 n-heptane (*nH_m_frac*), 0.305 toluene (*Tol_m_frac*) and 0.103 ethanol (*Eth_m_frac*);

  a) Alternatively, a custom composition can be defined by entering *def_surr_comp = no* (calculation with default surrogate composition is turned off), *specify_surr_comp = yes* (the composition of the surrogate shall be specified), followed by the values mass fractions of the four surrogate components: *iO_m_frac = value*, *nH_m_frac = value*, *Tol_m_frac = value*, *Eth_m_frac = value*;

  b) Another alternative is the estimation of the surrogate composition from the real fuel's properties, as discussed in Section 6.2. In this case, the gasoline RON (*RON = value*), ethanol content in volume percent (*V_ethanol* = *value*) as well as two arbiratray characteristics from the following list are needed: MON (*MON* = *value*), liquid density (*rho [kg/m^3]* = *value*), lower heating value (*H_u [MJ/kg]* = *value*), H/C ratio (*HC_ratio [-]* = *value*) or number of H, C and O atoms per molecule (*H_fuel (-)* = *value*, *C_fuel (-)* = *value*, *O_fuel (-)* = *value*). Additionally, the calculation of the surrogate composition has to be activated by entering *def_surr_comp* = *no* (calculation with default surrogate composition is turned off) and *specify_surr_properties* = *yes* (the specified real fuel properties of the surrogate shall be considered);

■ Consideration of the horizontal spark plug offset (which is an optional calibration parameter of the Entrainment model) and neglect of the corresponding value in vertical direction (constant value of 4 mm in the Entrainment model that should not be recalibrated): *ZK_hor_an* = *yes*, *ZK_ver_an* = *no*;

■ The high-temperature ignition integral always starts at 0.3: *Ik_NTZ = 0.3*;

■ Both integrals that are evaluated in the course of the auto-ignition prediction are not scaled: *C_NTZ = 1*, *C_HTZ = 1*;

■ The estimation of all unburnt mixture properties at the temperature of the boundary layer is deactivated by default: *eval_bl_calo* = *off*;

■ The modification of the unburnt temperature level is turned off by default: *temp_scale [permille]* = *0*. The unburnt temperature curve is modified throughout and the sign of the entered value defines if the level is reduced (-) or increased (+). The scaling of the unburnt temperature is performed in per mill and not by a constant value to ensure that the unburnt temperature ratio between the single operating points remains unaffected by significant temperature level changes (e.g. in case of an engine load variation), as this would lead to considerable knock prediction errors;

■ The normalization of the curve of the unburnt mass fraction in the thermal boundary layer is active: *norm_x_ub_grenz* = *yes*;

■ The curve of the unburnt mass fraction in the thermal boundary layer is additionally being smoothed by default with a factor of 0.0025: *bl_smoothing* = *yes*, *smoothing_factor* = *0.0025*;

■ The estimated value of the calibration parameter representing the experimental knock boundary can be entered prior to performing predictive simulations as follows: $x\_ub\_grenz\_sz = value$.

The knock model validation in Chapter 7 has been performed with the default set of parameter values listed here.

6.3.2 Workflow

The complete model workflow for the prediction of the knock limited spark advance is shown in Figure 6.7. The knock boundary estimation is based on the combustion simulation with a two-zone SI combustion model (and all involved models, e.g. laminar flame speed, calorific properties calculation, turbulence, wall heat transfer etc.), which yields the data of the unburnt mixture parameters, such as mean temperature, in-cylinder pressure and mixture composition (AFR, exhaust gas and injected water fractions). These values are then used for the prediction of auto-ignition with the two-stage approach proposed in Section 4.5.1. The integration process is performed over time. At each integration step, the value of the ignition delay for the corresponding ignition stage is evaluated as a function of the current boundary conditions in the unburnt mixture with the matching sub-model presented in Section 4.4. The integration is started at 90 °CA before firing top dead center or later, if at this point the inlet valves have not been closed yet. It should be remarked that the start-of-integration point does not influence the value of the low-temperature ignition integral in a meaningful way (see Figure 2.5), as long as it is before 20 °CAbFTDC (approximate value, stating a value that is always true is not possible, as it depends on the change of the boundary conditions in the unburnt mixture during the compression stroke). This is because prior to the late compression phase, the low unburnt temperature and in-cylinder pressure result in low-temperature ignition delay times that are relatively high, so that the integrand is very small and thus, the integral value does not change considerably.

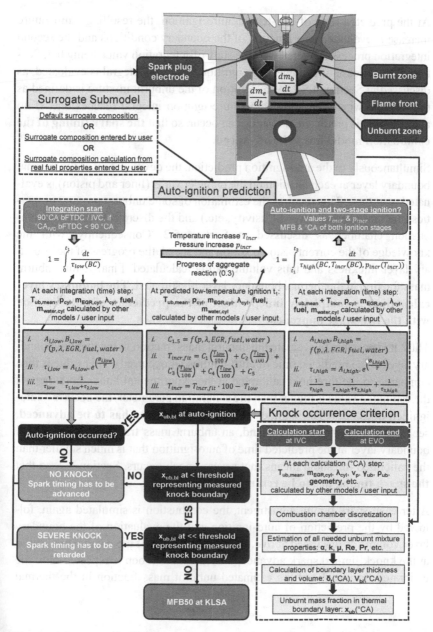

Figure 6.7: Knock limited spark advance estimation workflow with the newly developed knock model.

At the predicted time of low-temperature ignition, the resulting temperature increase is evaluated as a function of the boundary conditions and the second integration process is started with the start-of-integration value being 0.3, Section 4.5.1. In case of a single stage ignition, only one integral is evaluated, as discussed in Section 4.5.2. Auto-ignition of the unburnt mixture is defined as the location where the high-temperature ignition integral reaches one. If no auto-ignition is predicted, no knock can occur, so that **the spark timing of the combustion model has to be advanced**.

Simultaneously to the auto-ignition prediction, the development of the thermal boundary layer at each location on the cylinder walls (liner and piston) is evaluated over time. This requires the estimation of some unburnt mixture properties (thermal conductivity, diffusivity, etc.) and the discretization of the combustion chamber as discussed in Section 5.2. Consequently, with the knowledge of the current piston position and speed, the progress of the boundary layer thickness and thus volume can be calculated. Finally, the unburnt mass fraction in the thermal boundary layer is estimated. **The boundary layer calculation is completely independent of the prediction of auto-ignition with the two-stage approach.**

Thus, it is possible to evaluate the unburnt mass fraction in the thermal boundary layer at the predicted time of auto-ignition (if occurred). This value is then compared to the threshold calibrated at the experimental knock boundary. If the estimated value is bigger than the calibration parameter, the occurred auto-ignition does not result in knock and **the spark timing has to be advanced**, see Figure 5.5. On the other hand, an unburnt mass fraction in the thermal boundary layer at the predicted time of auto-ignition that is much smaller than the calibrated threshold means that severe knock occurs, Section 5.2, so that **the spark timing has to be retarded**, Figure 5.5.

After each spark timing adjustment, the combustion is simulated again, followed by the prediction of auto-ignition and the evaluation of the boundary layer volume as discussed above. For the estimation of the combustion center at the knock limited spark advance, this calculation loop is repeated until the calibrated threshold and the estimated unburnt mass fraction in the thermal boundary layer at the predicted time of auto-ignition coincide[16]. As discussed in Section 5.2, the knock limited spark advance in the simulation can be tuned with a simple controller that adjusts the spark timing of the combustion model based on the above reflections.

Because of the small time step during the calculation of the low- and high-temperature ignition integrals of 10^{-7} seconds, which has been selected in Section 4.5.1 as a good compromise between accuracy and performance, the computational cost of the knock model proposed in this work is relatively high for a model of this type. The integration process is performed until either combustion end or auto-ignition is reached. The development of the thermal boundary layer on the other hand is estimated in the °CA-domain and is much faster.

The computational cost of the knock model generally depends on the combustion duration, as longer combustion results in a higher number of values that have to be integrated. Furthermore, because of the constant integration step size in the time domain, the computational cost of the model is also significantly influenced by the engine speed. Thus, the worst case regarding the computational speed of the knock model is represented by an operating point at low engine speed and late combustion center position. In particular, the knock prediction takes approximately 66 milliseconds per engine cycle (about 40 % of the total cycle simulation duration) on a typical quad-core CPU for an operating point at the knock limited spark advance at 1500 min⁻¹ with an indicated mean pressure of 20 bar and a MFB50-point at 28 °CAaFTDC. If the engine speed is increased to 4000 min⁻¹ and the indicated mean pressure is reduced to 12 bar, the combustion center at the knock limited spark advance shifts to 12 °CAaFTDC, Figure 7.8, and the knock prediction for one engine cycle is completed within less than 50 milliseconds. Thus, the iterative calculation of the knock limited spark advance started at an arbitrary combustion center position and performed by varying the spark timing (or directly the MFB50-point) typically takes no more than 2 to 2.5 seconds.

---

[16] This task can be simplified by defining a tolerance range for the unburnt mass fraction in the thermal boundary layer at KLSA, which is similar to the definition of the experimental knock boundary as a knock frequency window.

6.3.3   Post-Processing

Because of the assumptions made in the course of the boundary layer volume model development in Section 5.2, the unburnt mass fraction in the layer before firing top dead center equals zero. At combustion end there is no unburnt mass left, so that the fraction in the boundary layer equals infinity. However, for simplicity reasons, this is not considered and the value is set to zero after the end of combustion. Thus, the theoretical maximum of the fraction is one and is reached at the point where the unburnt zone is completely within the thermal boundary layer (very late combustion phase).

Nevertheless, the estimated unburnt temperature level never equals exactly the one in a real combustion engine. This error influences the calculated boundary layer temperature and thus the values of the unburnt mixture properties that are needed for the layer development calculation, such as the thermal diffusivity and conductivity. Furthermore, the values of these parameters are yielded by empirical correlations (Equations 5.5 to 5.10) and are thus approximate. Additionally, the flame propagation causes a demolishment of the boundary layer at all locations of contact with the cylinder wall. The slight spark plug position offset in the Entrainment model leads to an unsymmetrical flame geometry, thus correctly representing the deviations from a perfectly spherical propagation that are always observed in reality. However, the real shape of the flame is still much more complex than the one the combustion model assumes.

Therefore, it is clear that the maximum of the modeled unburnt mass fraction in the thermal boundary layer will never equal one. Furthermore, investigations at various operating conditions have revealed that the mass fraction maxima can vary significantly, primarily with engine speed as shown in Figure 6.8, thus leading to a knock boundary prediction error. As the engine speed strongly influences the cycle duration in the time domain, the observed effect of this parameter presumably partially results from the change in the time available for the boundary layer development. Because the knock simulation is performed based on a constant absolute value of the unburnt mass fraction in the thermal boundary layer, it is clear that in the case of fraction maxima that are changing systematically with the operating conditions, e.g. decrease with engine speed, the value of the knock model calibration parameter has to be modified accordingly.

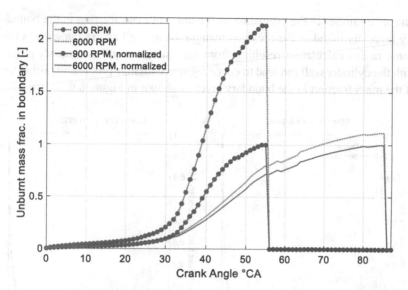

Figure 6.8:   Exemplary influence of engine speed on the maximum of the calculated unburnt mass fraction in the thermal boundary layer and effect of the implemented curve normalization.

Alternatively, the unburnt mass fraction in the thermal boundary layer can be normalized as shown in Equation 6.1. Thus, the fraction maxima always equal one and the knock model calibration parameter does not have to be modified. Figure 6.8 shows a comparison between the default and the normalized curves for two extreme cases revealing that the implemented curve normalization significantly influences the locations in °CA where a specified constant unburnt mass fraction in the boundary layer representing KLSA, e.g. 0.5, is reached and exceeded. The implemented normalization contributes to the very high accuracy of the knock boundary prediction over engine speed achieved with the new knock model, which will be demonstrated in Chapter 7.

$$x_{ub,bl,norm}(i) = \frac{x_{ub,bl}(i)}{max(x_{ub,bl})} \qquad \text{Eq. 6.1}$$

i calculation step index [-]

$x_{ub,bl,norm}$ normalized unburnt mass fraction in thermal boundary layer [-]

The uncertainties in the evaluation of the unburnt mass fraction in the boundary layer discussed in the previous paragraphs as well as inaccuracies in the flame radius calculation resulting from the estimation of the contact points with the cylinder wall can lead to curve regions with non-progressive behavior of the mass fraction in the boundary layer as shown in Figure 6.9.

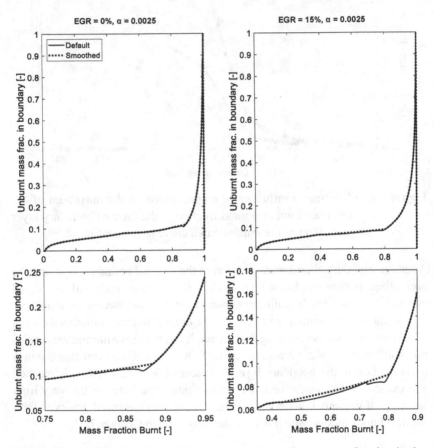

Figure 6.9: Default and smoothed curves of the unburnt mass fraction in the thermal boundary layer (top) with the smoothed curve regions zoomed in (bottom).

In this case, more than one MFB-points are characterized by the same unburnt mass fraction in the thermal boundary layer. Thus, if the value of the knock model calibration parameter $x_{ub,bl}$ is within the non-progressive curve region, it is unclear, which of the locations characterized by the same unburnt mass fraction in the boundary layer corresponds to the experimental knock boundary. Hence, in this case the knock limit cannot be reliably predicted.

Extensive investigations have revealed that non-progressive curve regions occur very rarely. Nevertheless, as it is clear that these are caused by inaccuracies in the various correlations and sub-models discussed above, it is justified to smoothen the curve of the unburnt mass fraction in the thermal boundary layer, yielding an always-progressive curve. However, as smoothing the curve means modifying it, it also causes a deviation from the originally estimated values that could negatively influence the accuracy of the knock boundary prediction. Thus, on the one hand, the unburnt mass fraction in the boundary layer has to allow the clear identification of the knock boundary, but an "over-smoothed", heavily modified curve will lead to knock prediction errors.

$$x_{ub,bl,smth}(i) = x_{ub,bl}(i-1) \cdot (1+\alpha) \qquad \text{Eq. 6.2}$$

$x_{ub,bl,smth}$ smoothed unburnt mass fraction in thermal boundary layer [-]

i calculation step index [-]

α smoothing factor [-]

To this end, a simple smoothing procedure has been implemented as shown in Equation 6.2. The curve smoothing is active, as long as the current value of the unburnt mass fraction in the boundary layer is smaller than the previous smoothed one. The smoothing factor does not change with the operating conditions. The calculation yields curves that are always progressive as shown in Figure 6.9 and allow the clear identification of the MFB-point corresponding to a specific unburnt mass fraction in the boundary layer.

Figure 6.10 shows that in case of a non-progressive unburnt mass fraction curve (here at 5 %, 10 % and 15 % EGR), the implemented smoothing procedure with the selected default factor of 0.0025 leads to a considerable increase in the accuracy of the knock boundary prediction. Higher smoothing factors however will be counterproductive, as these will result in modifications of the unburnt mass fraction curve that are too considerable.

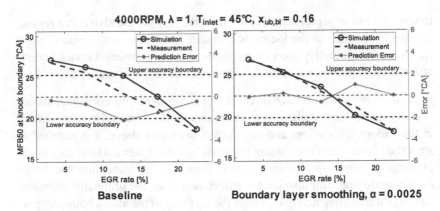

Figure 6.10: Effect of the proposed boundary layer curve smoothing on the knock prediction accuracy.

6.3.4 Model Outputs

Table 6.1 summarizes all outputs of the newly developed 0D/1D knock model. The parameters have been selected so that the user can gain a deep insight into the auto-ignition process as well as the occurrence of knock resulting from an occurred auto-ignition. The model yields if the mixture auto-ignites and when (expressed in °CA and MFB for each ignition stage in case of a two-stage ignition) as well as the progress of the ignition integrals and the corresponding values of the ignition delay times. Additionally, in case of a two-stage ignition, the temperature increase resulting from the low-temperature heat release as well as the corresponding pressure rise are yielded. A further model output is the progress of the unburnt mass fraction in the thermal boundary layer as well as the corresponding value at auto-ignition and if knock has occurred.

Finally, it should be remarked once again that the unburnt temperature level is the parameter with by far the most pronounced influence on the calibrated unburnt mass fraction in the thermal boundary layer at the time of auto-ignition, see Section 6.1. It also significantly affects the calculated °CA and MFB at auto-ignition and thus knock onset. The unburnt temperature level of averaged cycles used for the knock boundary prediction however is calculated by other simulation models. Hence, despite the physical background of the knock model calibration parameter $x_{ub,bl}$ and the time of auto-ignition expressed in °CA and MFB, the absolute values of these parameters yielded by the knock model are not meaningful and should not be interpreted. However, they can be helpful to identify potential problems, as for example a two-stage ignition occurring with a low-temperature heat release taking place before combustion start is a sign that the fuel selected for the knock boundary simulation, the wall heat transfer parameters and the cylinder wall temperature values should be checked, because the unburnt temperature level is probably too high.

Table 6.1:   Outputs of the newly developed 0D/1D knock model for one simulated engine cycle.

Mean Parameter Values for the simulated working cycle	Progress of Parameters over °CA
• Surrogate composition (component mass fractions)[17] • Properties of the calculated surrogate composition: RON, MON, LHV, liquid density, H/C ratio[18] • Auto-ignition occurred? • Two-stage ignition occurred? • Knock occurred? • Temperature (T_{incr}) and pressure (p_{incr}) increase resulting from the occurred low-temperature ignition[19] • °CA & MFB-point of the occurred low-temperature ignition[19] • °CA & MFB-point of the occurred auto-ignition[20] • Unburnt mass fraction in thermal boundary layer at the time of auto-ignition[20] • Calibration parameter value	• Low-temperature ignition integral • Low-temperature ignition delay • High-temperature ignition integral • High-temperature ignition delay • Progress of the unburnt mass fraction in thermal boundary layer

---

[17] Only in case the composition has been entered by the user or calculated from real fuel properties and thus differs from the default values representing the base fuel in Section 4.2.

[18] Only in case the surrogate composition has been estimated from properties of the real fuel, Sections 6.2 and 6.3.1.

[19] Only in case two-stage ignition occurred.

[20] Only in case auto-ignition occurred.

7 Knock Model Validation

In this chapter, the newly developed 0D/1D knock model is applied to available measurement data to assess its prediction accuracy and the results are discussed. The performance of the newly developed knock model has been evaluated extensively at various operating conditions on three different engines. The validation process involved the comparison of the measured MFB50-points at the knock boundary with the ones predicted by the knock model in the course of 0D/1D engine simulations, as the combustion center is decisive for the engine efficiency [75] [122]. For this purpose, all measured operating points were simulated with 0D/1D and three pressure analysis models of the investigated engine configurations. All simulation models involved (e.g. two-zone combustion, turbulence etc.) were calibrated so that they reproduce the experimental behavior of each engine as accurately as possible. Thus, the simulations yield temperature and pressure curves as well as heat release rates similar to those estimated from the test bench data.

The performed model validation is based completely on data yielded by predictive 0D/1D simulations. **The single knock model calibration parameter $x_{ub,bl}$ has been estimated at one operating point for each of the investigated engines respectively** as explained in Section 6.3.1 **and subsequently kept constant during the validation process.** It should be remarked that the knock prediction approach could also be directly applied to the available measured mean and single cycles. However, the model developer believes that such a validation procedure is neither expedient nor sensible, as the new knock model is supposed to enable the fully predictive simulation of the knock boundary, where the knock prediction is based on simulated and not experimentally obtained temperature and pressure curves. The experimental knock limited spark advance as well as the knock onset of measured single cycles on the other hand can be easily obtained as discussed in Sections 3.1 and 3.2, even without performing a detailed combustion analysis of the experimental data.

The main validation goal was to **ensure the model responds correctly to changes in the engine configuration and operating conditions represented by the direction of shift of the MFB50-points at the knock limited spark advance.** A further objective was to **achieve an error in the predicted MFB50-point at the knock boundary of 2 °CA or less** with the constant

© Springer Fachmedien Wiesbaden GmbH, part of Springer Nature 2019
A. Fandakov, *A Phenomenological Knock Model for the Development of Future Engine Concepts*, Wissenschaftliche Reihe Fahrzeugtechnik Universität Stuttgart, https://doi.org/10.1007/978-3-658-24875-8_7

engine-specific threshold for the unburnt mass fraction in the cool boundary layer calibrated at the experimental knock limit. Furthermore, because the knock model does not include any empirical sub-models that could contain engine-specific effects, **it is expected that the quality of knock boundary prediction does not vary considerably between different engines**, provided that the calibration quality of the rest of the simulation models is similar.

7.1   Engine 1

The first engine the knock model performance was evaluated on is the single-cylinder configuration presented in Section 3.1. The performance of the knock boundary prediction is demonstrated in Figure 7.1 to Figure 7.6. The investigated operating points include variations of various operating conditions, such as engine speed, external EGR rate, inlet temperature and the AFR. Changes in the tumble and engine coolant and oil temperatures at different EGR rates have been assessed too.

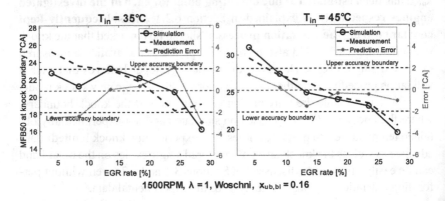

Figure 7.1:   Knock model validation, engine 1: EGR and inlet temperature at 1500 min⁻¹.

Overall, the prediction performance of the new knock model is very good and the effects of all operating conditions on the KLSA are predicted correctly. When assessing the accuracy of the new model, it should be considered that each measured point at the experimental knock limit has a slightly different knock frequency, because the boundary was defined as a knock frequency window, Section 3.1. Nevertheless, the simulated MFB50-points at the knock boundary are within the defined limits of 2 °CA. The high knock prediction quality with both wall heat transfer approaches demonstrated here enables the investigation of different combustion systems, engine configurations and operating conditions within a 0D/1D simulation environment as well as their influence on knock occurrence, burn duration and hence the indicated efficiency.

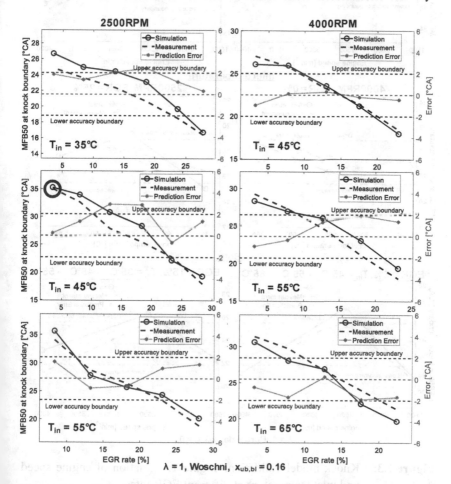

Figure 7.2: Knock model validation, engine 1: Engine speed, EGR and inlet temperature. The circle marks the operating point the knock model was calibrated at for the investigated engine.

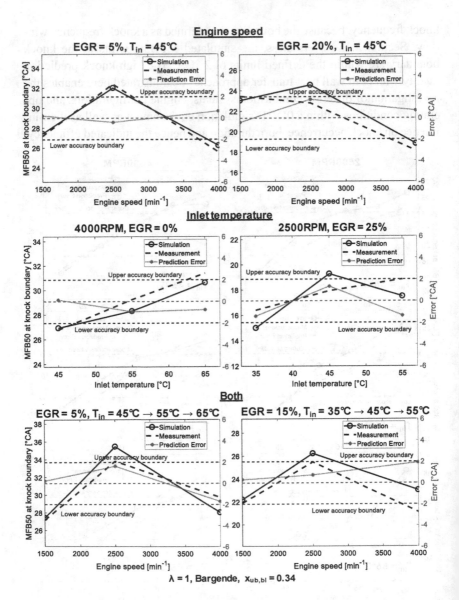

Figure 7.3: Knock model validation, engine 1: Variation of engine speed and inlet temperature at different EGR rates.

The measurement data generally show that the high heat capacity of the cooled exhaust gas allows for earlier MFB50-points at the knock boundary with external EGR, resulting in an overall higher indicated efficiency (the achievable efficiency increase depends on the engine load, Section 7.2), although the burn duration increases because of the exhaust gas' influence on the laminar flame speed. Additionally, Figure 7.5 demonstrates that EGR can be used to replace the mixture enrichment typically applied at full load, thus again resulting in a significant indicated efficiency gain.

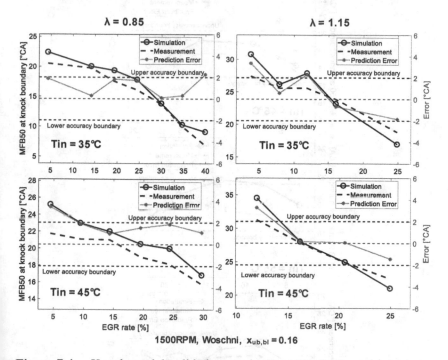

Figure 7.4: Knock model validation, engine 1: AFR, EGR, and inlet temperature.

Figure 7.6 shows that the new model correctly accounts for the in-cylinder turbulence level's effect on the knock behavior. Lower turbulence levels, here achieved by changing the position of the engine tumble flap, influence the knock occurrence significantly because they result in longer burn durations and thus lower peak in-cylinder pressures and unburnt temperatures. As these parameters also represent the knock model inputs, the new approach is able to capture the influence of the in-cylinder turbulence level on the knock boundary correctly. Hence, an explicit consideration of the in-cylinder turbulence level within the knock model (e.g. by adding a term for the influence of turbulence as proposed by Schmid in [135]) is not necessary.

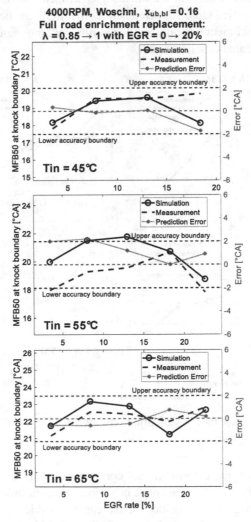

Figure 7.5:   Knock model validation, engine 1: Full load enrichment replacement with EGR (constant pre-turbine temperature) at different inlet temperatures.

Furthermore, the knock model correctly accounts for engine coolant and oil temperature variations, Figure 7.6. These cause a change in the cylinder wall temperature level that, as already discussed in detail in Section 6.1, typically influences the wall heat transfer and thus the unburnt temperature level.

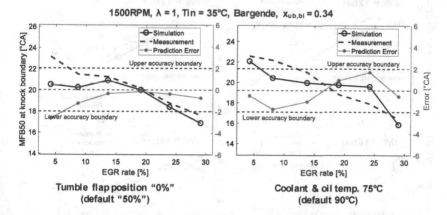

Figure 7.6: Knock model validation, engine 1: EGR rate variation at lower tumble and engine coolant / oil temperature levels.

Overall, it can be concluded that the quality of the predicted values is generally very high and that the newly developed knock model correctly reproduces all effects of the operating condition variations investigated in the course of the performed predictive 0D/1D simulations.

7.2 Engine 2

The second investigated engine is a single-cylinder configuration that differs significantly from engine 1, as it has different injection system and piston, a higher compression ratio as well as a tumble runner design, see Section 3.1 [51] [54]. The corresponding model validation results can be found in Figure 7.7 and Figure 7.8, where a load increase over EGR at different engine speeds is evaluated. As discussed in the previous section, the results show that generally, the high heat capacity of the cooled exhaust gas allows for earlier

MFB50-points at the knock boundary with EGR, although the exhaust gas decreases the laminar flame speed, thus resulting in slower combustion.

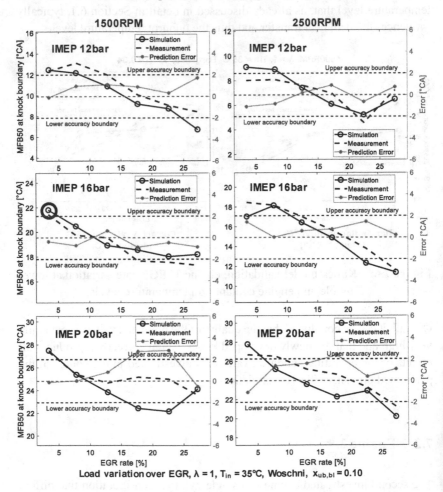

Figure 7.7: Knock model validation, engine 2: Engine speed, load, and EGR. The circle marks the operating point the knock model was calibrated at for the investigated engine.

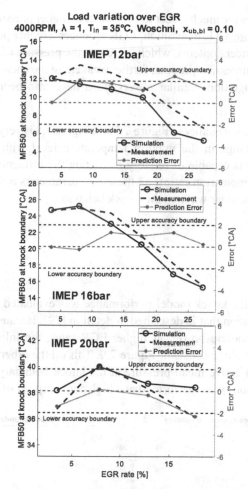

Figure 7.8: Knock model validation, engine 2: Engine load and EGR at 4000 min⁻¹.

The data in Figure 7.7 and Figure 7.8 further reveals the influence of engine speed and load on the knock mitigation effect of EGR. Higher engine loads generally result in later combustion at the knock boundary, so that the unburnt temperature level decreases. In the meantime, the required higher power output causes an in-cylinder fuel and air mass increase, so that the pressure level in the compression stroke as well as at top dead center rises. Thus, the suppression of knock by EGR is reduced, as the ignition delay time curves for different EGR rates are much closer to each other at higher pressures and lower temperatures, Figure 4.11. In addition, EGR produces a higher polytrophic exponent of the cylinder contents, which increases the pressure of the unburned gas during the compression process, further weakening the chemical influence of EGR on auto-ignition. Similar observations have also been recently reported in [145].

Overall, the prediction quality in Figure 7.7 and Figure 7.8 is very high, despite the significant change in the unburnt temperature level with engine load. Hence, the model is capable of reliably predicting the influence of different loads and thus compression ratios, as these two parameters affect the knock model inputs in a similar way, on the knock behavior.

7.3   Engine 3

The third engine the knock model performance was evaluated on is a three-cylinder configuration that is described in detail in [79]. Because of the EGR cooler's limited cooling power, at higher EGR rates, the inlet temperature could not be maintained constant, Figure 7.9. This on the other hand led to a decrease in the effective mean pressure at the knock boundary. The experimental investigations were performed at a high engine speed of 5500 min⁻¹.

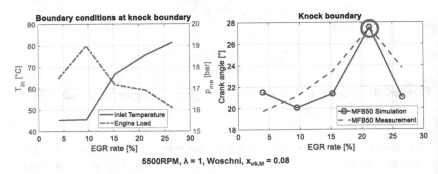

Figure 7.9:   Knock model validation, engine 3: EGR, inlet temperature, and engine load. The circle marks the operating point the knock model was calibrated at for the investigated engine.

Overall, the measurement series shown in Figure 7.9 contains significant simultaneous changes in a handful of the operating conditions. Despite the complex interrelations these result in, the performance of the knock model is very good, with both the direction of the shift and the absolute values of the MFB50-points at the knock boundary being predicted accurately by the newly developed model.

7.4 Conclusions

Overall, the performed model validation has demonstrated that **the newly developed approach for the 0D/1D simulation considers changes in the engine configuration and operating conditions correctly by accurately predicting the direction of the shift of the MFB50-points as well as their absolute values at the knock limited spark advance**. Moreover, **the model has been successfully validated on three different engines with no significant variations in the knock boundary prediction quality**. It has been demonstrated that an excellent knock simulation performance can be achieved with the mean values of the unburnt mixture's parameters being used as model inputs, even if the operating conditions change significantly. This is possible thanks to the **lack of any empirical sub-models, the novel two-stage auto-ignition prediction approach as well as the proposed cycle-individual phenomenological knock occurrence criterion** that takes cylinder geometry, flame propagation and the operating conditions into account. Finally, in the following the quality of the knock boundary simulation achieved with the new model shall be compared with the prediction performance of two commercial knock models commonly used today.

Figure 7.10 shows a comparison between the knock boundaries (represented by their MFB50-points) predicted with the newly developed knock model and a commercial approach (denoted as "OLD") proposed by Schmid in 2010 [125] [135] that is based on the knock model developed by Worret [168], Section 2.3.2.2. Prior to the performed evaluation, the two models have been calibrated at a single operating point and subsequently the values of their calibration parameters have been kept constant.

The increasing error over EGR of the commercial model can be attributed to the missing influence of exhaust gas on the mixture ignition delay [135]. The huge prediction inaccuracies over engine speed and load however result from the significant effect of these operating parameters on the two-stage ignition behavior of the unburnt mixture, Figure 4.3. The presence of exhaust gas on the other hand affects the low-temperature ignition only marginally.

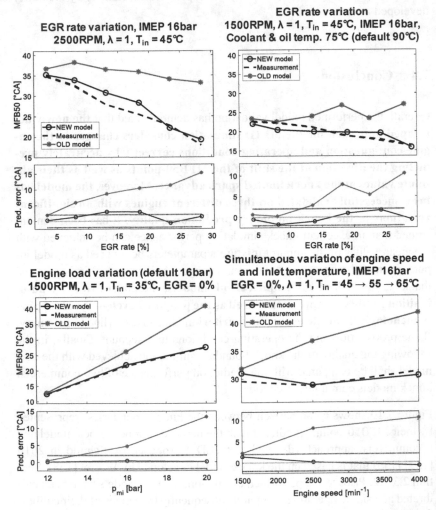

Figure 7.10: Knock boundary prediction with a nowadays commonly used commercial [135] and the newly developed knock models, engines 1 and 2.

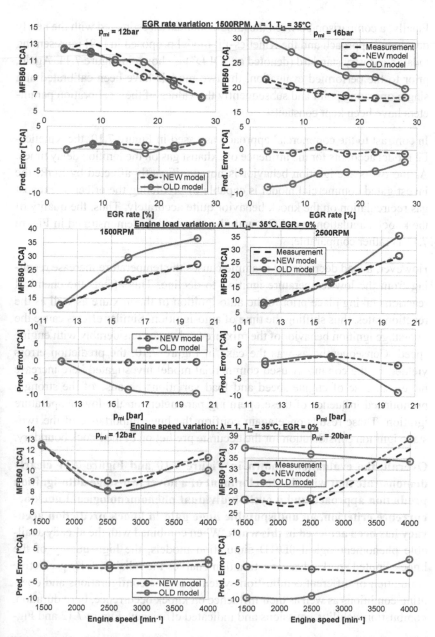

Figure 7.11: Knock boundary prediction with today's industry standard [69] and the newly developed knock model, engine 2.

Finally, a comparison between the knock boundaries estimated with the newly developed approach and the Kinetics-Fit model proposed in [69] representing today's industry standard (denoted as "OLD") is shown in Figure 7.11. Again, prior to the performed evaluation, the two models have been calibrated at a single operating point and subsequently the values of their calibration parameters have been kept constant.

In contrast to the commercial approach assessed in Figure 7.10, the Kinetics-Fit model accounts for the influence of exhaust gas on the ignition delay times. As the two-stage ignition behavior is not significantly affected by EGR, the investigated commercial model is capable of predicting the effect of exhaust gas recirculation on the knock behavior quite accurately. Thus, the quality of the knock simulation over EGR with the two models demonstrated in Figure 7.11 is rather comparable.

However, a load decrease for example causes the temperature increase resulting from the low-temperature ignition to decline. This affects the chemical reactions taking place prior to the auto-ignition of the mixture as well as the reaction rates, thus shifting the time of auto-ignition. Similar effects cause the two-stage ignition behavior of the mixture to change considerably with engine speed. Consequently, as was the case in Figure 7.10, the prediction errors yielded by the commonly used commercial model investigated here increase hugely in case of engine speed and load variations because of the strongly pronounced influence of these operating parameters on the low-temperature ignition. These results once again confirm the huge importance of the two-stage ignition consideration for the accurate prediction of the knock boundary.

Overall, the evaluations shown in Figure 7.10 and Figure 7.11 clearly demonstrate that with the development of a novel two-stage auto-ignition prediction approach and a cycle-individual phenomenological knock occurrence criterion, an undisputable huge gain in knock prediction accuracy has been achieved in this work. When combined with the already available phenomenological 0D/1D simulation models, the high quality of the knock limited spark advance estimation enables the reliable, fully predictive simulation of different combustion systems, engine configurations and operating conditions as well as their influence on knock occurrence, burn duration, combustion stability, emissions and indicated efficiency, Figure 7.12 and Figure 7.13. Thus, the new phenomenological knock model contributes substantially to the efficient development process of future spark ignition engine concepts within a 0D/1D simulation environment.

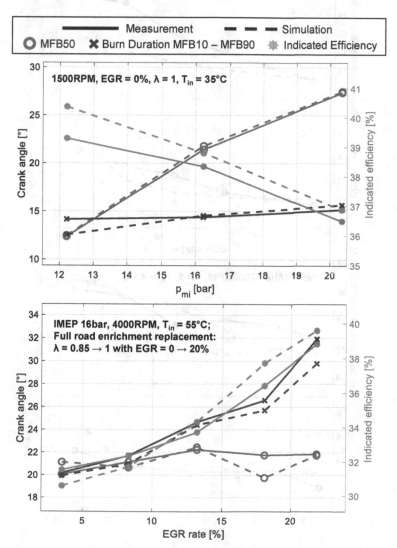

Figure 7.12: Simulated effect of full-load enrichment replacement by EGR and load influence on KLSA, indicated efficiency and burn duration, engines 1 and 2.

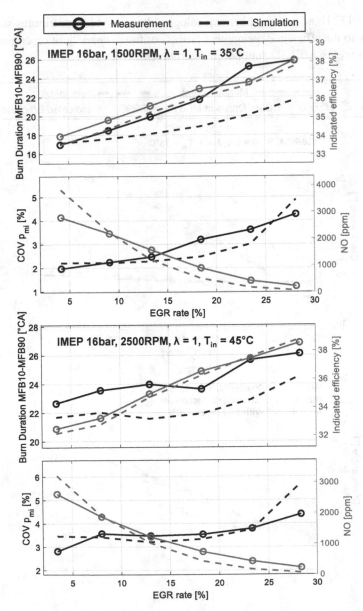

Figure 7.13: Simulation of the influence of low-pressure EGR on KLSA, indicated efficiency, burn duration, combustion stability and NO emissions at different engine speeds, engine 1.

8 Conclusions and Outlook

Overall, a lot of effort has been put into the development of a reliable 0D/1D knock model in the past decades. The huge number of researches that have worked and are still working on this topic as well as the regularly communicated poor knock prediction accuracy suggested that major changes in the simplified approaches for modeling the real chemical and physical processes are needed in order to improve the knock prediction performance. This required the better understanding of the processes that lead to knock in real engines.

In the present work, all tools available today have been employed for the formulation of a new, more appropriate phenomenological approach for the description of knock occurrence, resulting in a novel knock model for the development of future engine concepts within a 0D/1D simulation environment. The performed thermodynamic analysis and the experimental investigations have proven that the generally applied knock integral is not capable of accurately estimating the progress of the chemical reactions leading to auto-ignition in the unburnt mixture. Additionally, the commonly assumed constant end-of-integration and thus latest possible MFB-point where knock can occur is not feasible and leads to prediction errors. Hence, for the accurate simulation of the knock boundary within a 0D/1D simulation environment, both a new auto-ignition prediction approach and a cycle-individual knock occurrence criterion had to be developed.

The performed reaction kinetic simulations at in-cylinder conditions have revealed that the occurrence of two-stage ignition results in partially huge errors and hence poor quality of the auto-ignition prediction performed with the commonly used Livengood-Wu integral. Besides, an influence of engine speed, exhaust gas and surrogate composition on the behavior of the prediction performance was observed. For some single cycles, no auto-ignition was predicted by the phenomenological approach at all, although the detailed mechanism auto-ignited. Thus, it got obvious that the knock integral cannot account for the low-temperature ignition and its influence on the mixture's ignition delay. Livengood and Wu already surmised this circumstance in the course of their pioneer research in 1955. Hence, none of the available knock models for gasoline fuels based on the Livengood-Wu integral is fully predictive. The

© Springer Fachmedien Wiesbaden GmbH, part of Springer Nature 2019
A. Fandakov, *A Phenomenological Knock Model for the Development of
Future Engine Concepts*, Wissenschaftliche Reihe Fahrzeugtechnik Universität
Stuttgart, https://doi.org/10.1007/978-3-658-24875-8_8

general knock modeling approach had to be improved, in order to consider the occurrence of two-stage ignition in the unburnt mixture that results in knock.

Based on these findings, a new two-stage approach reproducing the auto-ignition behavior of the detailed mechanism at in-cylinder conditions was developed. The occurrence of each of the two ignition events is predicted by a single integral. The inputs of the two coupled integrals are the values of the ignition delay for the corresponding ignition stage as a function of the current boundary conditions. For this purpose, an enhanced three-zone approach for modeling the influence of various parameters (pressure, temperature, exhaust gas, air-fuel equivalence ratio, ethanol and water content as well as surrogate composition) on the auto-ignition delay times of the mixture was developed. Furthermore, models for the delay of the low-temperature ignition as well as the temperature increase resulting from the first ignition stage as a function of the boundary conditions were formulated. Finally, it was demonstrated that the novel auto-ignition model predicts the occurrence of two-stage ignition and considers the significant influence of low-temperature heat release on the mixture's auto-ignition behavior very accurately at various operating conditions.

Furthermore, as an auto-ignition does not necessarily result in knock, a knock occurrence criterion based on the unburnt mass fraction in the thermal boundary layer was proposed. It accounts for effects of the current operating conditions, as these influence both the boundary layer development and the flame propagation significantly, as well as for the flame propagation itself. Thus, the developed phenomenological approach yields a cycle-individual latest MFB-point where knock can occur. To this end, it is assumed that if the unburnt mass fraction in the boundary layer at the predicted time of auto-ignition is higher than a pre-defined threshold calibrated at the measured knock boundary, no knock can occur. The criterion contains no empirical measurement data fits and considers a handful of cylinder geometry parameters influencing the boundary layer, such as the top dead center clearance, bore, stroke, spark plug position, and the piston diameter. Thus, the new knock model accounts for differences in the engine geometry and has just one calibration parameter – the unburnt mass fraction in the thermal boundary layer at the time of auto-ignition calibrated at the experimental knock boundary – that is engine-specific and does not change with the operating conditions. Therefore, after a simple recalibration, the new, fully predictive knock model can be applied to different engines without any limitations.

Additionally, detailed investigations have shown that the proposed phenomenological knock occurrence criterion based on the unburnt mass fraction in the boundary layer yields accurate prediction results, even if the operating conditions are varied considerably and the calculations are performed with the mean unburnt temperature and the global AFR and exhaust mass fractions and. However, because of the significant influence of other simulation models on the knock model calibration parameter, its absolute value is not meaningful and should not be interpreted, despite its physical background. The mean unburnt temperature level was identified as the parameter with by far the most pronounced influence on the value of the calibrated unburnt mass fraction in the thermal boundary layer at the time of auto-ignition.

Nevertheless, further evaluations revealed that a similar prediction quality is achievable with different wall heat transfer approaches and cylinder wall temperature parametrizations, although these influence the mean unburnt temperature level significantly. In addition, it became apparent that the simulation of single cycles currently does not improve the knock boundary prediction accuracy and leads to a significant increase in computational time. With its high number of possible user inputs, the developed submodel for the estimation of the surrogate composition significantly simplifies the definition of the fuel for the knock boundary prediction as well as the investigation of the influence of fuel properties on the knock behavior.

The performed extensive model validation revealed that the newly developed knock model for the 0D/1D simulation considers changes in the engine configuration and operating conditions correctly by accurately predicting the direction of the shift of the MFB50-points as well as their absolute values at the knock limited spark advance. Moreover, the model has been successfully validated on three different engines with no significant variations in the knock boundary prediction performance. Finally, the comparison of the knock boundary simulation quality achieved with the new model and two available knock models commonly used today clearly demonstrated that with the development of a novel two-stage auto-ignition prediction approach and a cycle-individual phenomenological knock occurrence criterion, an undisputable huge gain in knock prediction accuracy has been achieved in this work. Thus, the new knock model shall contribute substantially to the efficient development process of future spark ignition engine concepts within a 0D/1D simulation environment. Additionally, the two-stage auto-ignition model proposed in this work can be incorporated into 3D CFD RANS simulations and employed for the prediction of local auto-ignition in the end-gas zone.

By coupling a phenomenological inhomogeneities model (as long as one is available) with a CCV simulation approach in the future, a further knock prediction accuracy gain is expected, as the combined models shall be capable of yielding the exact values of the knock model inputs at the location where knock is initiated. Additionally, the knowledge of these exact values is supposed to result in unburnt mass fractions in the thermal boundary layer that are always meaningful and less dependent on the unburnt temperature level. The knock modeling approach proposed in this work can be further expanded for the knock boundary simulation with fuels different from ethanol-doped gasoline, e.g. methane and synthetic fuels. This requires the development and validation of respective reaction kinetic models based on available ignition delay time measurements. Such an expansion has to account for the fact that not all fuel types suitable for the use in SI engines show a two-stage auto-ignition behavior. Additionally, as the fuel type can influence the combustion process significantly (e.g. because of considerable effects on the laminar flame speed and thus on the flame propagation), the appropriateness of the proposed knock occurrence criterion has to be examined in this case. Finally, the knock model accuracy could further benefit from modeling the effects of particles, e.g. from engine oil, that influence the auto-ignition process. Such a model extension requires the detailed reaction kinetic description of different lubricants as well as the accurate estimation of the engine oil fraction in the cylinder based on various factors, e.g. operating conditions and combustion chamber geometry.

Bibliography

[1] Adolph N., "Messung des Klopfens an Ottomotoren", Dissertation, Rheinisch Westfälische Technische Hochschule Aachen, 1983.

[2] Anderson, J. E., Kramer, U., Mueller, S. A., Wallington, T. J., "Octane Numbers of Ethanol- and Methanol-Gasoline Blends Estimated from Molar Concentrations," Energy Fuels, 24, 2010, pp. 6576-6585.

[3] Ando, H., Takemura, J., and Koujina, E., "A Knock Anticipating Strategy Basing on the Real-Time Combustion Mode Analysis," SAE Technical Paper 890882, 1989, doi: 10.4271/890882.

[4] Arrhenius, S.A., "On the Influence of Carbonic Acid in the Air Upon the Temperature of the Ground," Philosophical Magazine, Vol. 41, 1896, pp. 237-76.

[5] Arrhenius, S.A., "Über die Dissociationswärme und den Einfluß der Temperatur auf den Dissociationsgrad der Elektrolyte," Zeitschrift für physikalische Chemie Vol. 4, 1889, pp. 96-116.

[6] Arrigoni V. et al., "Quantitative Systems for Measuring Knock," Proc. Instn. Mech. Engrs, 1972.

[7] Auer, M., Wachtmeister, G., "Erstellung eines phänomenologischen Modells zur Vorausberechnung des Brennverlaufes von Gasmotoren", Final Report on FVV Project 874, Research Association for Combustion Engines (FVV) e.V., Frankfurt am Main, 2008.

[8] Bargende, M., "Ein Gleichungsansatz zur Berechnung der instationären Wandwärmeverluste im Hochdruckteil von Ottomotoren," Dissertation, Darmstadt, Technische Hochschule, 1991.

[9] Bargende, M., Burkhardt, C., Frommelt, A., "Besonderheiten der thermodynamischen Analyse von DE Ottomotoren," MTZ Motorentechnische Zeitschrift 62, 2001, pp. 56-68.

© Springer Fachmedien Wiesbaden GmbH, part of Springer Nature 2019
A. Fandakov, *A Phenomenological Knock Model for the Development of Future Engine Concepts*, Wissenschaftliche Reihe Fahrzeugtechnik Universität Stuttgart, https://doi.org/10.1007/978-3-658-24875-8

[10]   Bargende, M., Heinle, M., Berner, H.-J., "Some Useful Additions to
       Calculate the Wall Heat Losses in Real Cycle Simulations," 13. Con-
       ference „The Working Process of the Internal Combustion Engine,"
       Graz, 2011, pp. 45-63.

[11]   Barnard, J. A., Bradley, J., N., "Flame and Combustion," Second Edi-
       tion, Chapman and Hall, 1985.

[12]   Barros, S., Atkinson, W., and Piduru, N., "Extraction of Liquid Water
       from the Exhaust of a Diesel Engine," SAE Technical Paper 2015-01-
       2806, 2015, doi: 10.4271/2015-01-2806.

[13]   Barton, R., Lestz, S., and Duke, L., "Knock Intensity as a Function of
       Engine Rate of Pressure Change," SAE Technical Paper 700061,
       1970, doi: 10.4271/700061.

[14]   Battistoni, M.; Mariani, F.; Risi, F.; Poggiani, C., "Combustion CFD
       Modeling of a Spark Ignited Optical Access Engine Fueled with Gas-
       oline and Ethanol," Energy Procedia, 82, 2015, pp. 424-431, doi:
       10.1016/j.egypro.2015.11.829.

[15]   Beckers, A., Motortechnische Zeitschrift, 14, Nr. 12, 345, 1953.

[16]   Benson, G., Fletcher, E., Murphy, T., and Scherrer, H., "Knock (Det-
       onation) Control by Engine Combustion Chamber Shape," SAE Tech-
       nical Paper 830509, 1983, doi: 10.4271/830509.

[17]   Bossung, C., Bargende, M., Dingel, O., Grill, M., "A quasi-dimen-
       sional charge motion and turbulence model for engine process calcu-
       lations," 15th Stuttgart International Symposium, Springer Vieweg,
       2015, ISBN 978-3-658-08844-6.

[18]   Bougrine, S., Richard, S., Veynante, D., "Modelling and Simulation
       of the Combustion of Ethanol blended Fuels in a SI Engine using a 0D
       Coherent Flame Model," SAE Technical Paper 2009-24-0016, 2009,
       doi: 10.4271/2009-24-0016.

[19]   Bozza, F., De Bellis, V., Minarelli, F., Cacciatore, D., "Knock and
       Cycle by Cycle Analysis of a High Performance V12 Spark Ignition
       Engine. Part 2: 1D Combustion and Knock Modeling," SAE Int. J.
       Engines 8(5):2002-2011, 2015, doi: 10.4271/2015-24-2393.

[20] Brunt, M., Pond, C., and Biundo, J., "Gasoline Engine Knock Analysis using Cylinder Pressure Data," SAE Technical Paper 980896, 1998, doi: 10.4271/980896.

[21] Burgdorf, K. and Chomiak, J., "A New Knock Form - an Experimental Study," SAE Technical Paper 982589, 1998, doi: 10.4271/982589.

[22] Burgdorf, K. and Denbratt, I., "A Contribution to Knock Statistics," SAE Technical Paper 982475, 1998, doi: 10.4271/982475.

[23] Burgdorf, K., Denbratt, I., "Comparison of Cylinder Pressure Based Knock Detection Methods," SAE Technical Paper 972932, 1997, doi: 10.4271/972932.

[24] Burke, S. M., Burke, U., McDonagh, R. et al., "An experimental and mode-ling study of propene oxidation. Part 2: Ignition delay time and flame speed measurements," Combustion and Flame, Volume 162, Issue 2, 2015, pp. 296-314.

[25] Burluka, A., Liu, K., Sheppard, C., Smallbone, A. et al., "The Influence of Simulated Residual and NO Concentrations on Knock Onset for PRFs and Gasolines," SAE Technical Paper 2004-01-2998, 2004, doi: 10.4271/2004-01-2998.

[26] Cai, L. and Pitsch, H., "Optimized chemical mechanism for combustion of gasoline surrogate fuels," Combustion and Flame, 162, 2015, pp. 1623-1637.

[27] Cai, L., "Chemical Kinetic Mechanism Development and Optimization for Conventional and Alternative Fuels," Shaker, 2016, ISBN 978-3-8440-4611-3.

[28] Cai, L., Fandakov, A., Mally, M., Ramalingam, A., Minwegen, H., "Knock with EGR at full load," Final Report on FVV Project 6301, H1144, Research Association for Combustion Engines e.V. (FVV), Frankfurt am Main, 2017.

[29] Cai, L., Pitsch, H., Mohamed, S.Y.; Raman, Bugler, V.J.; Curran, H., Sarathy, S.M.: "Optimized reaction mechanism rate rules for ignition of normal alkanes," Combustion and Flame, 173, 2016.

[30]   Cavina, N., Corti, E., Minelli, G., Moro, D. et al., "Knock Indexes
        Normalization Methodologies," SAE Technical Paper 2006-01-2998,
        2006, https://doi.org/10.4271/2006-01-2998.

[31]   Checkel, M., Dale, J., "Computerized Knock Detection from Engine
        Pressure Records," SAE Technical Paper 860028, 1986, doi: 10.4271/
        860028.

[32]   Checkel, M., Dale, J., "Pressure Trace Knock Measurement in a Cur-
        rent S.I. Production Engine," SAE Technical Paper 890243, 1989, doi:
        10.4271/890243.

[33]   Checkel, M., Dale, J., "Testing a Third Derivative Knock Indicator on
        a Production Engine," SAE Technical Paper 861216, 1986, doi:
        10.4271/861216.

[34]   Chen, L., Li, T., Yin, T., Zheng, B., "A predictive model for knock
        onset in spark-ignition engines with cooled EGR," Energy Conversion
        and Management, 946-955, 2014, doi: 101016/jenconman201408002.

[35]   Chiodi, M., "An innovative 3D-CFD-Approach towards Virtual De-
        velopment of Internal Combustion Engines," Dissertation, University
        of Stuttgart, 2010.

[36]   Chun, K. and Heywood, J., "Characterization of Knock in a Spark-
        Ignition Engine," SAE Technical Paper 890156, 1989, doi: 10.4271/
        890156.

[37]   Chun, K. M., Heywood, J. B., Keck, J. C., "Prediction of knock oc-
        currence in a spark-ignition engine," Proc Combust Inst, 22 (1) (1989),
        pp. 455-463.

[38]   Cracknell, R. F., Prakash, A., Somers, K. P., Wang, C., "Impact of
        Detailed Fuel Chemistry on Knocking Behaviour in Engines," 5. In-
        ternational Conference on Knocking in Gasoline Engines, Springer,
        Cham, 2017, ISBN 978-3-319-69759-8

[39]   Csallner, P., "Eine Methode zur Vorausberechnung der Änderung des
        Brennverlustes von Ottomotoren bei geänderten Betriebsbedingun-
        gen", Dissertation, Technical University of Munich, 1981.

[40] Curran, H., Gaffuri, P., Pitz, W.J. and Westbrook, C.K., "A compre-
hensive modeling study of iso-octane oxidation," Combustion and
Flame, Volume 129, 2002, pp. 253-280.

[41] Dahnz, C., Han, K., Spicher, U., Magar, M. et al., "Investigations on
Pre-Ignition in Highly Supercharged SI Engines," SAE Int. J. Engines
3(1):214-224, 2010, https://doi.org/10.4271/2010-01-0355.

[42] David G. Goodwin, Harry K. Moffat, Raymond L. Speth., "Cantera:
An object- oriented software toolkit for chemical kinetics, thermodynamics, and transport processes," http://www.cantera.org, 2017, Version 2.3.0, doi: 10.5281/zenodo.170284.

[43] Davis, W., Smith, M., Malmberg, E., Bobbitt, J., "Comparison of In-
termediate-Combustion Products Formed in Engine with and without
Ignition," SAE Technical Paper 550262, 1955, doi: 10.4271/550262.

[44] Dechaux, J.C., Delfosse, L., "The negative temperature coefficient in
the C2 to C13 hydrocarbon oxidation. I. Morphological results", Combustion and Flame, 34, 1979, pp. 161-168.

[45] Douaud, A., Eyzat, P., "Four-Octane-Number Method for Predicting
the Anti-Knock Behavior of Fuels and Engines," SAE Technical Paper
780080, 1978, doi: 10.4271/780080.

[46] Downes, D., "Chemical and Physical Studies of Engine Knock," Six
Lectures on the Basic Combustion Processes, pp. 127-155, Ethyl Cor-
poration, Detroit, Mich., 1954.

[47] Downs D., Street J. C., Wheeler R. W. "Cool flame formation in a
motored engine," Fuel, 1953, 32, 279.

[48] Durst, B., Unterweger, G., Rubbert, S., "Thermodynamic effects of
water injection on Otto Cycle engines with different water injection
systems," 15. Conference „The Working Process of the Internal Combustion Engine," Graz, 2015, pp. 443-453.

[49] Eichlseder, H., Klüting, M., Piock, W., "Grundlagen und Technologien des Ottomotors," Wien: Springer, 2008, ISBN 978-3-211-25774-
6.

[50]   Elmqvist, C., Lindström, F., Ångström, H., Grandin, B. et al., "Optimizing Engine Concepts by Using a Simple Model for Knock Prediction," SAE Technical Paper 2003-01-3123, 2003, doi: 10.4271/2003-01-3123.

[51]   Fandakov, A., Bargende, M., Grill, M., Mally, M., Kulzer, A. C., "A new model for predicting the knock boundary with EGR at full load," 10. MTZ Conference on the Charge Cycle in Combustion Engines, Stuttgart, 2017.

[52]   Fandakov, A., Grill, M., Bargende, M., Kulzer, A. C., "Investigation of thermodynamic and chemical influences on knock for the working process calculation," 17th Stuttgart International Symposium, Springer Vieweg, 2017, ISBN 978-3-658-16987-9.

[53]   Fandakov, A., Grill, M., Bargende, M., Kulzer, A., "Two-Stage Ignition Occurrence in the End Gas and Modeling Its Influence on Engine Knock," SAE Int. J. Engines 10(4):2017, doi: 10.4271/2017-24-0001.

[54]   Fandakov, A., Mally, M., Cai, L. et al., "Development of a Model for Predicting the Knock Boundary in Consideration of Cooled Exhaust Gas Recirculation at Full Load," 5. International Conference on Knocking in Gasoline Engines, Springer, Cham, 2017, ISBN 978-3-319-69759-8.

[55]   Ferraro, C., Marzano, M., and Nuccio, P., "Knock-Limit Measurement in High-Speed S. l. Engines," SAE Technical Paper 850127, 1985, doi: 10.4271/850127.

[56]   Franzke, B., Morcinkowski, B., Adomeit, P., Hoppe, P., Esposito, S., "Potenziale der 1D-/ 3D-Kopplung in der Brennverfahrensentwicklung," 10. MTZ Conference on the Charge Cycle in Combustion Engines, Stuttgart, 2017

[57]   Franzke, D., "Beitrag zur Ermittlung eines Klopfkriteriums der ottomotorischen Verbrennung und zur Vorausberechnung der Klopfgrenze," Ph.D. thesis, Technical University of Munich, 1991.

[58] Fuiorescu, D., Radu, B., "A Proposed Relation for Knock Auto-Igni-
      tion Induction Period Evaluation in a LPG Fueled SI Engine," SAE
      Technical Paper 2010-01-1455, 2010, doi: 10.4271/2010-01-1455.

[59] Galloni, E., Fontana, G., Staccone, S., "Numerical and experimental
      characterization of knock occurrence in a turbo-charged spark-ignition
      engine," Energy Conversion and Management, 85, 2014, pp. 417-424,
      doi: 10.1016/j.enconman.2014.05.054

[60] Gauthier, B. M.; Davidson, D. F.; Hanson, R. K., "Shock tube deter-
      mination of ignition delay times in full-blend and surrogate fuel mix-
      tures," Combustion and Flame, 139 (4), 2004, pp. 300-311, doi:
      10.1016/j.combustflame.2004.08.015

[61] Ghojel, J., I., "Review of the development and applications of the
      Wiebe function: a tribute to the contribution of Ivan Wiebe to engine
      research, " International Journal of Engine Research, Vol. 11, 2010,
      pp. 297-312.

[62] Glassman, I., Yetter, R., A., "Combustion," 4th Edition, Elsevier Aca-
      demic Press, 2008, ISBN: 978-0-12-088573-2.

[63] Green, R.M., Pitz, R.M., Westbrook, C. K., "The Autoignition of iso-
      Butane in a Knocking Spark Ignition Engine," SAE Paper 870169,
      1987.

[64] Griffiths, J. F., "Reduced Kinetic Models and their Application to
      Practical Combustion Systems," Prog. Energy Combust. Sci., 1995,
      Vol. 21, pp. 25 - 107.

[65] Grill, M., "Objektorientierte Prozessrechnung von Verbrennungsmo-
      toren," Dissertation, University of Stuttgart, 2006.

[66] Grill, M., Bargende, M., "The Development of an Highly Modular De-
      signed Zero-Dimensional Engine Process Calculation Code," SAE Int.
      J. Engines 3(1):1-11, 2010, doi: 10.4271/2010-01-0149.

[67] Grill, M., Billinger, T., Bargende, M., "Quasi-Dimensional Modeling
      of Spark Ignition Engine Combustion with Variable Valve Train,"
      SAE-Paper 2006-01-1107, 2006.

[68] Grill, M., Chiodi, M., Berner, H. J., Bargende, M., "Calculating the thermodynamic properties of burnt gas and vapor fuel for user-defined fuels," MTZ worldwide, 68(5), pp. 30-35, 2007.

[69] GT-POWER User's Manual, GT-Suite Version 2017, Official Build 1, Gamma Technologies, August 2017.

[70] Haghgooie, M., "Effects of Fuel Octane Number and Inlet Air Temperature on Knock Characteristics of a Single Cylinder Engine," SAE Technical Paper 902134, 1990, doi: 10.4271/902134.

[71] Hann, S., Urban, L., Grill, M., and Bargende, M., "Influence of Binary CNG Substitute Composition on the Prediction of Burn Rate, Engine Knock and Cycle-to-Cycle Variations," SAE Int. J. Engines 10(2): 501-511, 2017, doi: 10.4271/2017-01-0518.

[72] Heinrich, C., Dörksen, H., Esch, A., Krämer, K., "Gasoline Water Direct Injection (GWDI) as a Key Feature for Future Gasoline Engines," 5. International Conference on Knocking in Gasoline Engines, Springer, Cham, 2017, ISBN 978-3-319-69759-8.

[73] Hermann, I., Glahn, C., Kluin, M., Paroll, M., Gumprich, W., "Water Injection for Gasoline Engines - Quo Vadis?," 5. International Conference on Knocking in Gasoline Engines, Springer, Cham, 2017, ISBN 978-3-319-69759-8.

[74] Hernández, J., Lapuerta, M., Sanz-Argent, J., "Autoignition prediction capability of the Livengood-Wu correlation applied to fuels of commercial interest," International Journal of Engine Research, Vol. 15, Number 7, 2014, pp. 817-829.

[75] Heywood J. B., "Internal combustion engine fundamentals," McGraw-Hill, Inc., USA, 1988, ISBN 978-0070286375.

[76] Hockel, K., "Untersuchung zur Laststeuerung beim Ottomotor," Dissertation, Technical University of Munich, 1982.

[77] Hoepke, B., Jannsen, S., Kasseris, E., Cheng, W., "EGR Effects on Boosted SI Engine Operation and Knock Integral Correlation," SAE Int. J. Engines 5(2):547-559, 2012, doi: 10.4271/2012-01-0707.

[78] Hoppe, F., Thewes, M., Seibel, J., Balazs, A. et al., "Evaluation of the Potential of Water Injection for Gasoline Engines," SAE Int. J. Engines 10(5):2017, doi: 10.4271/2017-24-0149.

[79] Hoppe, P., Lehrheuer, B., Pischinger, S., "Downsizing with Biofuels II," Final Report, FVV Spring Conference, Research Association for Combustion Engines (FVV) e.V., Bad Neuenahr, 2015.

[80] Hu, H., Keck, J., "Autoignition of Adiabatically Compressed Combustible Gas Mixtures," SAE Technical Paper 872110, 1987, doi: 10.4271/872110.

[81] Hunger, M., Böcking, T., Walther, U., Günther, M., Freisinger, N., Karl, G., "Potential of Direct Water Injection to Reduce Knocking and Increase the Efficiency of Gasoline Engines," 5. International Conference on Knocking in Gasoline Engines, Springer, Cham, 2017, ISBN 978-3-319-69759-8.

[82] Jakob, M., Pischinger, S., Adomeit, P., Brunn, A., Ewald, J.: "Effect of Intake Port Design on the Flow Field Stability of a Gasoline DI Engine," SAE Technical Paper 2011-01-1284, 2011.

[83] Käppner, C., Gonzalez, N. G., Drückhammer, J., Lange, H., Fritzsche, J., Henn, M., "On board water recovery for water injection in high efficiency gasoline engines," 17th Stuttgart International Symposium, Stuttgart, 2017, ISBN 978-3-658-16987-9.

[84] Kasseris, E., Heywood, J. B., "Charge cooling effects on knock limits in SIDI engines using gasoline/ethanol blends: part 2-effective octane numbers," SAE Inter J Fuels Lubricants, 2012, pp. 2:844-854.

[85] Keskin, M., T., "Modell zur Vorhersage der Brennrate in der Betriebsart kontrollierte Benzinselbstzündung," Wissenschaftliche Reihe Fahrzeugtechnik Universität Stuttgart, Springer Vieweg, 2016, doi: 10.1007/978-3-658-15065-5.

[86] Kim, K., Ghandhi, J., "Preliminary Results from a Simplified Approach to Modeling the Distribution of Engine Knock," SAE Technical Paper 2012-32-0004, 2012, doi: 10.4271/2012-32-0004.

[87]   Klimstra, J., "The Knock Severity Index – A Proposal for a Knock
       Classification Method," SAE Technical Paper 841335, 1984, doi:
       10.4271/841335.

[88]   Konig, G. and Sheppard, C., "End Gas Autoignition and Knock in a
       Spark Ignition Engine," SAE Technical Paper 902135, 1990, doi:
       10.4271/902135.

[89]   König, G., Maly, R., Bradley, D., Lau, A. et al., "Role of Exothermic
       Centres on Knock Initiation and Knock Damage," SAE Technical Pa-
       per 902136, 1990, doi: 10.4271/902136.

[90]   Kožuch, P.: "Ein phänomenologisches Modell zur kombinierten
       Stickoxid- und Rußberechnung bei direkteinspritzenden Dieselmoto-
       ren, " Dissertation, University of Stuttgart, 2004.

[91]   Kucheba, M., Bray, K., Ikonomou, E., and Cosman, A., "Calculations
       and Measurements of the Temperature Field in a Motored Engine and
       Their Implications for Knock," SAE Technical Paper 890844, 1989,
       https://doi.org/10.4271/890844.

[92]   Kukkadapu, G., Kumar, K., Sung, C.J., Mehl, M. and Pitz, W.J., "Au-
       toignition of gasoline and its surrogates in a rapid compression ma-
       chine", Proceed-ings of the Combustion Institute, Volume 34, Issue 1,
       2013, pp. 345-352.

[93]   Kuratle, R. and Märki, B., "Influencing Parameters and Error Sources
       During Indication on Internal Combustion Engines," SAE Technical
       Paper 920233, 1992, doi: 10.4271/920233.

[94]   Lafossas, F., Castagne, M., Dumas, J., and Henriot, S., "Development
       and Validation of a Knock Model in Spark Ignition Engines Using a
       CFD code," SAE Technical Paper 2002-01-2701, 2002, doi: 10.4271/
       2002-01-2701.

[95]   Lagarias, J. C., Reeds, J. A., Wright, M. H., Wright, P. E., "Conver-
       gence Properties of the Nelder-Mead Simplex Method in Low Dimen-
       sions," SIAM Journal of Optimization, Vol. 9, Number 1, 1998, pp.
       112-147.

[96] Lee, C., Vranckx, S., Heufer, K., Khomik, S., Uygun, Y., Olivier, H., Fenandes, R., "On the Chemical Kinetics of Ethanol Oxidation: Shock Tube, Rapid Compression Machine and Detailed Modeling Study," Zeitschrift für Physikalische Chemie, Volume 226, Issue 1, 2012, pp. 1-28.

[97] Lee, K., Yoon, M., Sunwoo, M., "A study on pegging methods for noisy cylinder pressure signal," Control Engineering Practice, Volume 16, Issue 8, 2008, pp. 922-929.

[98] Lee, S., Bae, C., Prucka, R., Fernandes, G. et al., "Quantification of Thermal Shock in a Piezoelectric Pressure Transducer," SAE Technical Paper 2005-01-2092, 2005, doi: 10.4271/2005-01-2092.

[99] Lee, W. and Schaefer, H., "Analysis of Local Pressures, Surface Temperatures and Engine Damages under Knock Conditions," SAE Technical Paper 830508, 1983, doi: 10.4271/830508.

[100] Leppard, W., "Individual-Cylinder Knock Occurence and Intensity in Multicylinder Engines," SAE Technical Paper 820074, 1982, doi: 10.4271/820074.

[101] Liang, L., Reitz, R. D., Yi, J., Iyer, C., "A *G*-equation Combustion Model Incorporating Detailed Chemical Kinetics for PFI/DI SI Engine Simulations," Sixteenth International Multidimensional Engine Modeling User's Group Meeting at the SAE Congress, 2006.

[102] Liang, L., Reitz, R., Iyer, C., and Yi, J., "Modeling Knock in Spark-Ignition Engines Using a G-equation Combustion Model Incorporating Detailed Chemical Kinetics," SAE Technical Paper 2007-01-0165, 2007, doi: 10.4271/2007-01-0165.

[103] Liebl, J., "Der Antrieb von morgen 2017: Hybride und elektrische Antriebssysteme 11. Internationale MTZ-Fachtagung Zukunftsantriebe," Wiesbaden: Springer, 2017, ISBN 978-3-658-19224-2.

[104] Linse, D., Kleemann, A., Hasse, C., "Probability density function approach coupled with detailed chemical kinetics for the prediction of knock in turbocharged direct injection spark ignition engines," Combust Flame, 161, 2014, pp. 997-1014.

[105] Livengood, J. C., Wu, P. C., "Correlation of autoignition phenomena in internal combustion engines and rapid compression machines," Symp. Int. Combust, 1955, 5: 347-356.

[106] Lyford-Pike, E. J., Heywood, J. B., "Thermal boundary layer thickness in the cylinder of a spark-ignition engine," Int. J. Heat Mass Transfer, 27 (10) (1984), pp. 1873-1878.

[107] Macpherson, J. H., Jr., Ph.D. Thesis, Stanford University, June 1946.

[108] Maiwald, O., "Experimentelle Untersuchungen und mathematische Modellierung von Verbrennungsprozessen in Motoren mit homogener Selbstzündung," Dissertation, University of Karlsruhe, 2005.

[109] Mansouri, S. H., Heywood, J. B., "Correlations for the viscosity and Prandtl number of hydrocarbon-air combustion products," Combust. Sci. Technol. 23, pp. 251-256, 1980.

[110] MATLAB 2015b/2016b/2017a, The MathWorks Inc., Natick, Massachusetts, 2015 – 2017.

[111] Merker, G., Teichmann, R., "Grundlagen Verbrennungsmotoren: Funktionsweise, Simulation, Messtechnik," 7. Auflage, Wiesbaden: Springer Vieweg, 2014, ISBN 978-3-658-03195-4.

[112] Minetti, R., Ribaucour, M., Carlier, M., Sochet, L. R., "Autoignition Delays of a Series of Linear and Branched Chain Alkanes in the Intermediate Range of Temperature", Combust. Sci. and Tech., 1996, Vols. 113 - 114, pp. 179 - 192.

[113] Mittal, V., Revier, B., Heywood, J., "Phenomena that Determine Knock Onset in Spark-Ignition Engines," SAE Technical Paper 2007-01-0007, 2007, doi: 10.4271/2007-01-0007.

[114] Morgan, N., Smallbone, A., Bhave, A., Kraft, M. et al., "Mapping surrogate gasoline compositions into RON/MON space," Combustion and Flame, 157, 2010, pp. 1122-1131.

[115] Moses E., Yarin A.L., Bar-Yoseph P., "On knocking prediction in spark ignition engines," Combustion and Flame, 101 (3), 1995, pp. 239-261.

[116] Müller, H., Bertling, H., "Programmierte Auswertung von Druckver-läufen in Ottomotoren " VDI Fortschrittsberichte Reihe 6, Nr. 30, Düsseldorf, 1971.

[117] Netzer, C., Seidel, L., Pasternak, M., Klauer, C. et al., "Engine Knock Prediction and Evaluation Based on Detonation Theory Using a Quasi-Dimensional Stochastic Reactor Model," SAE Technical Paper 2017-01-0538, 2017, doi: 10.4271/2017-01-0538.

[118] Ogink, R., "Computer Modeling of HCCI Combustion," Ph.D. Thesis, Chalmers University of Technology, 2004.

[119] Pan, J., Zhao, P., Law, C. K., Wei, H., "A predictive Livengood–Wu correlation for two-stage ignition," International Journal of Engine Research, Vol 17, Issue 8, 2016, pp. 825-835.

[120] Pepiot-Desjardins, P., Pitsch, H. "An efficient error-propagation-based reduction method for large chemical kinet-ic mechanisms," Combustion and Flame, Volume 154, 2008.

[121] Pflaum, W., Mollenhauer, K., "Wärmeübergang in der Verbrennungs-kraftmaschine," Springer-Verlag, 1977, ISBN 978-3-7091-8453-0.

[122] Pischinger, R., Klell, M., Sams, T., "Thermodynamik der Verbren-nungskraftmaschine," 3. Auflage, Springer-Verlag Wien, 2009, doi: 10.1007/978-3-211-99277-7.

[123] Pomraning, E., Richards, K., and Senecal, P., "Modeling Turbulent Combustion Using a RANS Model, Detailed Chemistry, and Adaptive Mesh Refinement," SAE Technical Paper 2014-01-1116, 2014, doi: 10.4271/2014-01-1116.

[124] Puzinauskas, P., "Examination of Methods Used to Characterize Engine Knock," SAE Technical Paper 920808, 1992, https://doi.org/10.4271/920808.

[125] Research Institute of Automotive Engineering and Vehicle Engines Stuttgart (FKFS), FKFS UserCylinder Compendium, Version 2.5.3, 2017.

[126] Rether, D., "Modell zur Vorhersage der Brennrate bei homogener und teilhomogener Dieselverbrennung," Dissertation, University of Stuttgart, 2012.

[127] Richard, S., Font, G., Le Berr, F., Grasset, O. et al., "On the Use of System Simulation to Explore the Potential of Innovative Combustion Systems: Methodology and Application to Highly Downsized SI Engines Running with Ethanol-Gasoline Blends," SAE Technical Paper 2011-01-0408, 2011, doi: 10.4271/2011-01-0408.

[128] Riess, M., Benz, A., Wöbke, M., Sens, M., "Einlassseitige Ventilhubstrategien zur Turbulenzgenerierung," MTZ, 74 (7-8), 2013.

[129] Rothe, M., Heidenreich, T., Spicher, U., and Schubert, A., "Knock Behavior of SI-Engines: Thermodynamic Analysis of Knock Onset Locations and Knock Intensities," SAE Technical Paper 2006-01-0225, 2006, https://doi.org/10.4271/2006-01-0225.

[130] Scharlipp, S., "Untersuchung des Klopfverhaltens methanbasierter Kraftstoffe," Wissenschaftliche Reihe Fahrzeugtechnik Universität Stuttgart, Springer Vieweg, 2017, doi: 10.1007/978-3-658-20205-7.

[131] Scherer, F. M., "Ladeluftkühlung durch Abgaswärmenutzung – ihr Einfluss auf die Abgasemission," Dissertation, Technical University of Berlin, 2014.

[132] Schiessl, R., Maas, U., "Analysis of Endgas Temperature Inhomogeneities in an SI Engine by Laser-Induced Fluorescence," Combustion and Flame, 133, 2003.

[133] Schießl, R., Schubert, A., and Maas, U., "Temperature Fluctuations in the Unburned Mixture: Indirect Visualisation Based on LIF and Numerical Simulations," SAE Technical Paper 2006-01-3338, 2006, doi: 10.4271/2006-01-3338.

[134] Schmid, A., Grill, M., Berner, H. J., Bargende, M., "Transient simulation with scavenging in the turbo spark-ignition engine," MTZ worldwide 71.11 (2010), pp. 10-15.

[135] Schmid, A., Grill, M., Berner, H.-J., Bargende, M., "A new Approach for SI-Knock Prediction," 3. International Conference on Knocking in Gasoline Engines, Berlin, 2010.

[136] Schmidt, F.A.F., "Verbrennungsmotoren, Thermodynamische und versuchsmässige Grundlage unter besonderer Berücksichtigung der Flugmotoren," p. 310, Berlin, Springer-Verlag, 1945.

[137] Schnaubelt, S., "Numerische Analyse des Selbstzündverhaltens einzelner Brennstofftropfen," Ph.D. Thesis, University of Bremen, 2005.

[138] Sens, M., Guenther, M., Hunger, M., Mueller, J. et al., "Achieving the Max - Potential from a Variable Compression Ratio and Early Intake Valve Closure Strategy by Combination with a Long Stroke Engine Layout," SAE Technical Paper 2017-24-0155, 2017, doi: 10.4271/ 2017-24-0155.

[139] Siokos, K., Koli, R. et al., "Assessment of Cooled Low Pressure EGR in a Turbocharged Direct Injection Gasoline Engine," SAE Int. J. Engines 8(4):1535-1543, 2015, doi: 10.4271/ 2015-01-1253.

[140] Song, H. and Song, H., "Knock Prediction of Two-Stage Ignition Fuels with Modified Livengood-Wu Integration Model by Cool Flame Elimination Method," SAE Technical Paper 2016-01-2294, 2016, doi: 10.4271/2016-01-2294.

[141] Stenlåås, O., Einewall, P., Egnell, R., and Johansson, B., "Measurement of Knock and Ion Current in a Spark Ignition Engine with and without NO Addition to the Intake Air," SAE Technical Paper 2003-01-0639, 2003, https://doi.org/10.4271/2003-01-0639.

[142] Steurs, K. F. H. M., "Cycle-resolved analysis and modeling of knock in a homogeneous charge spark ignition engine fueled by ethanol and iso-octane," Ph.D. Thesis, ETH Zurich, 2014.

[143] Sung, C.-J., Curran, H.J., "Using rapid compression machines for chemical kinetics studies," Progress in Energy and Combustion Science, Volume 44, 2014, pp. 1-18.

[144] Syed, I., Mukherjee, A., Naber, J., "Numerical Simulation of Autoignition of Gasoline-Ethanol/Air Mixtures under Different Conditions of Pressure, Temperature, Dilution, and Equivalence Ratio.,"
SAE Technical Paper 2011-01-0341, 2011, doi: 10.4271/2011-01-
0341.

[145] Szybist, J., Wagnon, S., Splitter, D., Pitz, W. et al., "The Reduced Ef-
fectiveness of EGR to Mitigate Knock at High Loads in Boosted SI
Engines," SAE Int. J. Engines 10(5):2017, doi: 10.4271/2017-24-
0061.

[146] Tabaczynski, R., Ferguson, C., Radhakrishnan, K., "A Turbulent Entrainment Model for Spark-Ignition Engine Combustion," SAE Technical Paper 770647, 1977, doi: 10.4271/770647.

[147] Tanaka, S., Ayala, F., Keck, J. C., Heywood, J. B., "Two-stage ignition in HCCI combustion and HCCI control by fuels and additives,"
Combustion and Flame, 132, 2003, pp. 219-239.

[148] Thewess, M., Baumgarten, H., Scharf, J., Hoppe, F., "Water Injection
– High Power and High Efficiency Combined," 25th Aachen Collo-
quium Automobile and Engine Technology, FEV Aachen, 2016.

[149] Thöne, H. J., "Untersuchung von Einflußgrößen auf das Klopfen von
Ottomotoren unter besonderer Beachtung der internen Abgasrückfüh-
rung," Dissertation, Rheinisch Westfälische Technische Hochschule
Aachen, 1994.

[150] Turns, S. "An Introduction to Combustion: Concepts and Applications," 3rd edition, McGraw Hill Higher Education, 2011, ISBN-10:
0073380199.

[151] Urban, L., Grill, M., Hann, S., Bargende, M., "Simulation of Autoignition, Knock and Combustion for Methane-Based Fuels," SAE Technical Paper 2017-01-2186, 2017.

[152] Valtadoros, T., Wong, V., Heywood, J., "Engine Knock Characteristics at the Audible Level," SAE Technical Paper 910567, 1991, doi:
10.4271/910567.

[153] Van Basshuysen, R., "Ottomotor mit Direkteinspritzung," 3. Auflage, Wiesbaden: Springer, 2013, ISBN 978-3-658-01407-0.

[154] Vancoillie, J., Sileghem, L., Verhelst, S., "Development and Validation of a Knock Prediction Model for Methanol-Fuelled SI Engines," SAE Technical Paper 2013-01-1312, 2013, doi: 10.4271/2013-01-1312.

[155] Vibe, I., I., "Brennverlauf und Kreisprozeß von Verbrennungsmotoren," VEB Verlag Technik, Berlin, 1970.

[156] Wang, Z., Liu, H., Reitz, R. D., "Knocking combustion in spark-ignition engines," Prog Energy Combust Sci, 61 (2017), pp. 78-112.

[157] Wang, Z., Wang, Y., Reitz, R. D., "Pressure Oscillation and Chemical Kinetics Coupling during Knock Processes in Gasoline Engine Combustion," Energy & Fuels, 26(12), 2012, pp. 7107-7119, doi: 10.1021/ef301472g.

[158] Warnatz, J., Mass, U., Dibble, R. W., "Combustion: Physical and Chemical Fundamentals, Modeling and Simulation, Experiments, Pollutant Formation," Springer Berlin Heidelberg, 2006.

[159] Wayne, W., Clark, N., Atkinson, C., "Numerical Prediction of Knock in a Bi-Fuel Engine," SAE Technical Paper 982533, 1998, doi: 10.4271/982533.

[160] Weisser, G. A., "Modelling of Combustion and Nitric Oxide Formation for Medium-Speed DI Diesel Engines: A Comparative Evaluation of Zero- and Three-Dimensional Approaches," Ph.D. Thesis, Swiss Federal Institute of Technology in Zurich (ETHZ), 2006.

[161] Wenig, M., "Simulation der ottomotorischen Zyklenschwankungen," Dissertation, University of Stuttgart, 2013.

[162] Wenig, M., Grill, M., Bargende, M., "A New Approach for Modeling Cycle-to-Cycle Variations within the Framework of a Real Working-Process Simulation," SAE Int. J. Engines 6(2):2013, doi: 10.4271/2013-01-1315.

[163] Westbrook, C. K., Pitz, W. J., "Detailed Kinetic Modeling of Autoignition Chemistry," SAE Paper 872107, 1987.

[164] Westbrook, C. K., Pitz, W. J., Leppard, W. R., "The Autoignition Chemistry of Paraffinic Fuels and Pro–Knock and Anti–Knock Additives: A Detailed Chemical Kinetic Study," SAE Paper 912314, 1991.

[165] Wheeler, R. W., "Abnormal Combustion Effects on Economy," J. C. Hilliard and G. S. Springer (eds.), Fuel Economy in Road Vehicles Powered by Spark-Ignition Engines, chap. 6, pp. 225-276, Plenum Press, 1984.

[166] Winklhofer, E., Philipp, H., Kapus, P., Piock, W. F., "Anomale Verbrennungseffekte in Ottomotoren," Haus der Technik Kongress," Tagung Optische Indizierung, Essen, 2004.

[167] Witt, A., "Analyse der thermodynamischen Verluste eines Ottomotors unter den Randbedingungen variabler Steuerzeiten," Dissertation, Graz University of Technology, 1999.

[168] Worret, R., Bernhardt, S., Schwarz, F., Spicher, U., "Application of Different Cylinder Pressure Based Knock Detection Methods in Spark Ignition Engines," SAE Technical Paper 2002-01-1668, 2002, doi: 10.4271/2002-01-1668.

[169] Woschni, G., "A Universally Applicable Equation for the Instantaneous Heat Transfer Coefficient in the Internal Combustion Engine," SAE Technical Paper 670931, 1967, doi: 10.4271/670931.

[170] Wurms, R., Budack, R., Grigo, M. et al., "Der neue Audi 2.0l Motor mit innovativem Rightsizing - ein weiterer Meilenstein der TFSI-Technologie," 36. International Vienna Motor Symposium, 2015.

[171] Yokomori, T., "Super-Lean Burn Technology for High Thermal Efficiency SI Engine," FVV Autumn Conference Proceedings (R580), Research Association for Combustion Engines (FVV) e.V., Leipzig, 2017.

[172] Zeldovich Y.B., "Regime Classification of an Exothermic Reaction with Nonuniformal Initial Conditions," Combustion and Flame, Volume 39, 1980, pp. 211-214.

[173] Zhang, K., Banyon, C., Bugler, J., Curran, H.J., Rodriguez, A., Herbinet, O., Battin-Leclerc, F., B'Chir, C. and Heufer, K.A., "An updated experimental and kinetic modeling study of n-heptane oxidation," Combustion and Flame, Volume 172, 2016, pp. 116-135.

[174] Zhen, X., Wang, Y., Xu, S. et al., "The engine knock analysis - An overview," Applied Energy, vol. 92, pp. 628-636, 2012.

[175] Zhen, X., Wang, Y., Zhu, Y., "Study of knock in a high compression ratio SI methanol engine using LES with detailed chemical kinetics," Energy Convers Manage, 2013, 75:523-31.

Appendix

The modeling approaches for the high- and low-temperature ignition delays as well as the temperature increase resulting from the first ignition stage have already been discussed in detail in Section 4.4. The following sections present all coefficients and the equations for calculating the main model parameters $A_{i,high}$, $B_{i,high}$, $A_{i,low}$, $B_{i,low}$ and $C_{1..5}$ as a function of the boundary conditions.

A1.   High-Temperature Ignition Delay Model

All model coefficients of the high-temperature ignition delay model (Equations 4.13 and 4.14), as well as the equations for calculating the main model parameters $A_{i,high}$ and $B_{i,high}$ as a function of the boundary conditions and surrogate composition (component mass fractions x) can be found in Table A.1. In order to increase the model accuracy by capturing the partially significant changes in the influences of the boundary conditions with pressure, all model parameters were modeled as a function of the current pressure p.

Table A.1:   High-temperature ignition delay model coefficients and equations for the calculation of the model parameters $A_{i,high}$ and $B_{i,high}$.

$$\tau_{i,high} = A_{i,high} \cdot e^{\left(\frac{B_{i,high}}{T}\right)}$$

$$A_{i,high} = exp\left(S_{lam} \cdot (\lambda - 1)^2 + P_{lam} \cdot (\lambda - 1) + S_{EGR} \cdot \left(\frac{m_{exh.gas}}{m_{cyl,total}} \cdot 100\right)^2 + P_{EGR} \cdot \right.$$

$$\frac{m_{exh.gas}}{m_{cyl,total}} \cdot 100 + S_{tol} \cdot \left(\frac{x_{Tol}}{x_{Iso}+x_{Hep}} - 0.514\right)^2 + P_{tol} \cdot \left(\frac{x_{Tol}}{x_{Iso}+x_{Hep}} - 0.514\right) + S_{iO} \cdot$$

$$\left(\frac{x_{Iso}}{x_{Iso}+x_{Hep}} - 0.223\right)^2 + P_{iO} \cdot \left(\frac{x_{Iso}}{x_{Iso}+x_{Hep}} - 0.223\right) + S_{etha} \cdot \left(\frac{x_{Eth}}{x_{Iso}+x_{Hep}+x_{Tol}} - 0.115\right)^2 +$$

$$\left. P_{etha} \cdot \left(\frac{x_{Eth}}{x_{Iso}+x_{Hep}+x_{Tol}} - 0.115\right) + U_A \right)$$

© Springer Fachmedien Wiesbaden GmbH, part of Springer Nature 2019
A. Fandakov, *A Phenomenological Knock Model for the Development of Future Engine Concepts*, Wissenschaftliche Reihe Fahrzeugtechnik Universität Stuttgart, https://doi.org/10.1007/978-3-658-24875-8

$$B_{i,high} = Q_{lam} \cdot (\lambda - 1)^2 + R_{lam} \cdot (\lambda - 1) + Q_{EGR} \cdot \left(\frac{m_{exh.gas}}{m_{cyl,total}} \cdot 100\right)^2 + R_{EGR} \cdot$$

$$\frac{m_{exh.gas}}{m_{cyl,total}} \cdot 100 + Q_{tol} \cdot \left(\frac{x_{Tol}}{x_{Iso}+x_{Hep}} - 0.514\right)^2 + R_{tol} \cdot \left(\frac{x_{Tol}}{x_{Iso}+x_{Hep}} - 0.514\right) + Q_{iO} \cdot$$

$$\left(\frac{x_{Iso}}{x_{Iso}+x_{Hep}} - 0.223\right)^2 + R_{iO} \cdot \left(\frac{x_{Iso}}{x_{Iso}+x_{Hep}} - 0.223\right) + Q_{etha} \cdot \left(\frac{x_{Eth}}{x_{Iso}+x_{Hep}+x_{Tol}} -$$

$$0.115\right)^2 + R_{etha} \cdot \left(\frac{x_{Eth}}{x_{Iso}+x_{Hep}+x_{Tol}} - 0.115\right) + U_B$$

$\tau_{1,high}$

Parameter	Function coefficients			Parameter	Function coefficients		
$S_{lam} =$ poly2(p)[A1]	-5.18E-5	9.75E-3	-6.21E-1	$Q_{lam} =$ poly2(p)	2.58E-2	-5.06E+0	2.88E+2
$P_{lam} =$ poly2(p)	-1.31E-4	2.44E-2	-1.48E+0	$R_{lam} =$ poly2(p)	7.73E-2	-1.42E+1	9.92E+2
$S_{EGR} =$ poly2(p)	1.97E-8	-8.29E-6	5.06E-4	$Q_{EGR} =$ poly2(p)	-9.13E-6	4.50E-3	-2.62E-1
$P_{EGR} =$ poly2(p)	2.21E-6	-3.10E-4	1.15E-2	$R_{EGR} =$ poly2(p)	-1.29E-3	1.76E-1	-4.06E+0
$S_{tol} =$ poly2(p)	7.85E-5	-9.94E-3	4.04E-1	$Q_{tol} =$ poly2(p)	-4.74E-2	5.98E+0	-2.30E+2
$P_{tol} =$ poly2(p)	-2.10E-4	2.81E-2	-2.08E+0	$R_{tol} =$ poly2(p)	1.24E-1	-1.65E+1	1.42E+3
$S_{iO} =$ poly2(p)	1.99E-4	-3.22E-2	-1.53E+0	$Q_{iO} =$ poly2(p)	-1.28E-1	2.09E+1	1.61E+3
$P_{iO} =$ poly2(p)	-2.72E-4	5.17E-2	3.05E-1	$R_{iO} =$ poly2(p)	1.70E-1	-3.26E+1	-1.61E+3
$S_{etha} =$ poly2(p)	5.11E-4	-7.89E-2	3.29E+0	$Q_{etha} =$ poly2(p)	-3.45E-1	5.39E+1	-2.28E+3
$P_{etha} =$ poly2(p)	-4.79E-4	9.24E-2	-7.39E+0	$R_{etha} =$ poly2(p)	2.92E-1	-5.66E+1	5.14E+3
$U_A =$ exp2(p)[A2]	3.4E+1	3.9E-4	6.5E+0	6.6E-3 / $U_B =$ exp2(p)	-6.9E+0	2.6E-2	1.6E+4	4.8E-4

$\tau_{2,high}$

Parameter	Function coefficients			Parameter	Function coefficients		
$S_{lam} =$ poly2(p)	-3.35E-5	6.80E-3	-2.46E+0	$Q_{lam} =$ poly2(p)	-1.66E-2	3.17E+0	1.30E+3
$P_{lam} =$ poly2(p)	-1.45E-5	1.90E-2	-2.37E+0	$R_{lam} =$ poly2(p)	-1.83E-2	-1.04E+1	2.85E+3
$S_{EGR} =$ poly2(p)	-9.21E-8	6.78E-6	-1.90E-4	$Q_{EGR} =$ poly2(p)	8.43E-5	-6.32E-3	4.10E-1
$P_{EGR} =$ poly2(p)	6.01E-6	-7.55E-4	5.89E-2	$R_{EGR} =$ poly2(p)	-4.51E-3	4.66E-1	-1.43E+1
$S_{tol} =$ poly2(p)	2.07E-4	-2.88E-2	2.90E-1	$Q_{tol} =$ poly2(p)	-1.61E-1	2.24E+1	-3.09E+2

P_{tol} = poly2(p)	-5.48E-4	8.12E-2	-2.34E+0	R_{tol} = poly2(p)	4.13E-1	-5.91E+1	2.29E+3
S_{iO} = poly2(p)	2.19E-4	-4.66E-2	4.72E+0	Q_{iO} = poly2(p)	-1.01E-1	2.20E+1	-2.68E+3
P_{iO} = poly2(p)	-4.02E-4	8.99E-2	-5.84E+0	R_{iO} = poly2(p)	2.16E-1	-5.05E+1	2.00E+3
S_{etha} = poly2(p)	5.55E-4	-7.85E-0	8.39E+0	Q_{etha} = poly2(p)	-4.16E-1	5.30E+1	-6.56E+3
P_{etha} = poly2(p)	-2.62E-4	7.16E-2	-1.12E+1	R_{etha} = poly2(p)	1.20E-1	-3.84E+1	9.26E+3
U_A = power2(p)[A3]	-9.63E+0	2.16E-1	2.04E+1	U_B = power2(p)	3.41E+2	5.97E-1	-6.88E+3

$$\tau_{3,high}$$

Parameter	Function coefficients				Parameter	Function coefficients			
S_{lam} = poly2(p)	9.35E-7	-5.74E-3	4.97E-1		Q_{lam} = poly2(p)	6.50E-3	5.04E+0	-7.03E+2	
P_{lam} = poly2(p)	3.43E-5	1.40E-2	-1.35E+0		R_{lam} = poly2(p)	-8.50E-2	-6.57E+0	1.80E+3	
S_{EGR} = poly2(p)	-1.02E-7	1.09E-5	-5.54E-5		Q_{EGR} = poly2(p)	1.24E-4	-1.26E-2	1.86E-1	
P_{EGR} = poly2(p)	4.12E-6	-1.63E-4	3.98E-2		R_{EGR} = poly2(p)	-5.04E-3	1.84E-1	-2.99E+1	
S_{tol} = poly2(p)	1.14E-4	-2.09E-2	-1.13E-1		Q_{tol} = poly2(p)	-1.21E-1	2.31E+1	2.60E+0	
P_{tol} = poly2(p)	-2.53E-4	5.14E-2	5.15E-1		R_{tol} = poly2(p)	2.66E-1	-5.54E+1	-7.76E+1	
S_{iO} = poly2(p)	1.05E-4	-7.17E-3	-4.76E-1		Q_{iO} = poly2(p)	-1.13E-1	6.82E+0	7.85E+2	
P_{iO} = poly2(p)	-3.55E-5	-1.33E-2	1.84E+0		R_{iO} = poly2(p)	5.70E-2	1.27E+1	-2.39E+3	
S_{etha} = poly2(p)	4.59E-4	-4.52E-2	8.82E+0		Q_{etha} = poly2(p)	-4.18E-1	2.31E+1	-7.75E+3	
P_{etha} = poly2(p)	7.79E-5	-4.78E-3	-6.99E+0		R_{etha} = poly2(p)	-1.74E-1	2.86E+1	6.20E+3	
U_A = exp2(p)	-2.3E+1	1.2E-3	2.9E+0	-9.5E-3	U_B = exp2(p)	2.1E+3	-8.0E-2	1.3E+4	2.2E-3

[A1] poly2 denotes a second order polynomial function with three coefficients $p_{1..3}$, so that $y = p_1 \cdot x^2 + p_2 \cdot x + p_3$.

[A2] exp2 denotes two coupled exponential functions with four coefficients $p_{1..4}$, so that $y = p_1 \cdot e^{p_2 \cdot x} + p_3 \cdot e^{p_4 \cdot x}$.

[A3] power2 denotes a power function with three coefficients $p_{1..3}$, so that $y = p_1 \cdot x^{p_2} + p_3$.

A2. Low-Temperature Ignition Delay Model

All model coefficients of the low-temperature ignition delay model (Equations 4.15 and 4.16), as well as the equations for calculating the main model parameters $A_{i,low}$ and $B_{i,low}$ as a function of the boundary conditions and surrogate composition (component mass fractions x) can be found in Table A.2. In order to increase the model accuracy by capturing the partially significant changes in the influences of the boundary conditions with pressure, all model parameters were modeled as a function of the current pressure p.

Table A.2: Low-temperature ignition delay model coefficients and equations for the calculation of the model parameters $A_{i,low}$ and $B_{i,low}$.

$$\tau_{i,low} = A_{i,low} \cdot e^{\left(\frac{B_{i,low}}{T}\right)}$$

$$A_{i,low} = exp\left(S_{lam} \cdot (\lambda - 1)^2 + P_{lam} \cdot (\lambda - 1) + S_{EGR} \cdot \left(\frac{m_{exh.gas}}{m_{cyl.total}} \cdot 100\right)^2 + P_{EGR} \cdot \right.$$

$$\frac{m_{exh.gas}}{m_{cyl.total}} \cdot 100 + S_{tol} \cdot \left(\frac{x_{Tol}}{x_{Iso}+x_{Hep}} - 0.514\right)^2 + P_{tol} \cdot \left(\frac{x_{Tol}}{x_{Iso}+x_{Hep}} - 0.514\right) + S_{iO} \cdot$$

$$\left(\frac{x_{Iso}}{x_{Iso}+x_{Hep}} - 0.223\right)^2 + P_{iO} \cdot \left(\frac{x_{Iso}}{x_{Iso}+x_{Hep}} - 0.223\right) + S_{etha} \cdot \left(\frac{x_{Eth}}{x_{Iso}+x_{Hep}+x_{Tol}} - 0.115\right)^2 +$$

$$\left. P_{etha} \cdot \left(\frac{x_{Eth}}{x_{Iso}+x_{Hep}+x_{Tol}} - 0.115\right) + U_A \right)$$

$$B_{i,low} = Q_{lam} \cdot (\lambda - 1)^2 + R_{lam} \cdot (\lambda - 1) + Q_{EGR} \cdot \left(\frac{m_{exh.gas}}{m_{cyl.total}} \cdot 100\right)^2 + R_{EGR} \cdot$$

$$\frac{m_{exh.gas}}{m_{cyl.total}} \cdot 100 + Q_{tol} \cdot \left(\frac{x_{Tol}}{x_{Iso}+x_{Hep}} - 0.514\right)^2 + R_{tol} \cdot \left(\frac{x_{Tol}}{x_{Iso}+x_{Hep}} - 0.514\right) + Q_{iO} \cdot$$

$$\left(\frac{x_{Iso}}{x_{Iso}+x_{Hep}} - 0.223\right)^2 + R_{iO} \cdot \left(\frac{x_{Iso}}{x_{Iso}+x_{Hep}} - 0.223\right) + Q_{etha} \cdot \left(\frac{x_{Eth}}{x_{Iso}+x_{Hep}+x_{Tol}} - \right.$$

$$\left. 0.115\right)^2 + R_{etha} \cdot \left(\frac{x_{Eth}}{x_{Iso}+x_{Hep}+x_{Tol}} - 0.115\right) + U_B$$

$\tau_{1,low}$					
Parameter	**Function coefficients**		**Parameter**	**Function coefficients**	
S_{lam} = poly2(p)	-1.54E-4	3.11E-2	-1.96E+0	Q_{lam} = poly2(p)	9.96E-2	-1.97E+1	1.15E+3
P_{lam} = poly2(p)	3.25E-4	-4.28E-2	1.52E+0	R_{lam} = poly2(p)	-2.08E-1	2.73E+1	-8.10E+2
S_{EGR} = poly2(p)	-1.90E-7	2.36E-5	-1.87E-4	Q_{EGR} = poly2(p)	1.26E-4	-1.58E-2	1.68E-1
P_{EGR} = poly2(p)	1.10E-05	-1.44E-3	3.91E-2	R_{EGR} = poly2(p)	-6.92E-3	8.92E-1	-2.08E+1

S_{tol} = poly2(p)	-6.98E-5	-5.18E-3	1.52E+0	Q_{tol} = poly2(p)	6.16E-2	6.93E-1	-8.53E+2
P_{tol} = poly2(p)	1.93E-4	5.86E-4	-2.68E+0	R_{tol} = poly2(p)	-1.50E-1	3.11E+0	1.73E+3
S_{iO} = poly2(p)	3.46E-5	-4.24E-2	4.58E-1	Q_{iO} = poly2(p)	1.04E-2	2.28E+1	4.69E+2
P_{iO} = poly2(p)	-2.11E-4	6.65E-2	-2.13E+0	R_{iO} = poly2(p)	1.01E-1	-3.75E+1	-1.99E+2
S_{etha} = poly2(p)	4.18E-4	-1.66E-1	1.45E+1	Q_{etha} = poly2(p)	-1.28E-1	8.28E+1	-8.52E+3
P_{etha} = poly2(p)	-1.40E-4	4.26E-2	-3.00E+0	R_{etha} = poly2(p)	8.61E-2	-2.86E+1	2.60E+3
U_A = power2(p)	1.50E+1	-2.66E-1	-3.32E+1	U_B = power2(p)	-7.85E+3	-2.35E-1	1.90E+4

$$\tau_{2,low}$$

Parameter	Function coefficients			Parameter	Function coefficients		
S_{lam} = poly2(p)	7.29E-4	-1.58E-2	-7.84E+0	Q_{lam} = poly2(p)	-7.12E-1	3.06E+1	5.79E+3
P_{lam} = poly2(p)	1.52E-3	-2.15E-1	8.88E+0	R_{lam} = poly2(p)	-1.26E+0	1.70E+2	-6.52E+3
S_{EGR} = poly2(p)	-9.80E-7	1.23E-4	-1.14E-3	Q_{EGR} = poly2(p)	8.67E-4	-1.09E-1	1.14E+0
P_{EGR} = poly2(p)	4.85E-5	-5.81E-3	1.09E-1	R_{EGR} = poly2(p)	-4.19E-2	4.91E+0	-7.03E+1
S_{tol} = poly2(p)	-5.50E-4	4.45E-2	1.66E+0	Q_{tol} = poly2(p)	5.50E-1	-4.91E+1	-1.15E+3
P_{tol} = poly2(p)	1.16E-3	-1.13E-1	-9.63E-1	R_{tol} = poly2(p)	-1.10E+0	1.11E+2	7.90E+2
S_{iO} = poly2(p)	-2.00E-3	1.97E-1	2.08E+0	Q_{iO} = poly2(p)	1.84E+0	-1.81E+2	-1.49E+3
P_{iO} = poly2(p)	6.04E-4	1.33E-2	-5.64E+0	R_{iO} = poly2(p)	-7.03E-1	1.59E+1	2.58E+3
S_{etha} = poly2(p)	-1.04E-3	-9.67E-2	2.77E+1	Q_{etha} = poly2(p)	1.53E+0	-2.23E+1	-2.05E+4
P_{etha} = poly2(p)	-6.29E-5	2.00E-2	1.16E+0	R_{etha} = poly2(p)	5.56E-2	-2.02E+1	3.58E+1
U_A = power2(p)	1.50E+2	-3.60E-2	-1.40E+2	U_B = power2(p)	1.64E+4	1.10E-1	-2.36E+4

A3. Model for Temperature Increase Resulting from Low-Temperature Ignition

All model coefficients of the temperature increase model (equations 4.17 and 4.18), as well as the equations for calculating the main model parameters $C_{1..5}$ as a function of the boundary conditions and surrogate composition (component mass fractions x) can be found in Table A.3. In order to increase the model accuracy by capturing the partially significant changes in the influences of the boundary conditions with pressure, all model parameters were modeled as a function of the current pressure p.

Table A.3: Temperature increase model coefficients and equations for the calculation of the model parameters $C_{1..5}$.

$$T_{incr,fit} = C_1 \left(\frac{T_{low}}{100}\right)^4 + C_2 \left(\frac{T_{low}}{100}\right)^3 + C_3 \left(\frac{T_{low}}{100}\right)^2 + C_4 \left(\frac{T_{low}}{100}\right)^1 + C_5$$

$$C_{1..5} = S_{lam} \cdot (\lambda - 1)^2 + P_{lam} \cdot (\lambda - 1) + S_{EGR} \cdot \left(\frac{m_{exh.gas}}{m_{cyl,total}} \cdot 100\right)^2 + P_{EGR} \cdot \frac{m_{exh.gas}}{m_{cyl,total}} \cdot$$
$$100 + S_{tol} \cdot \left(\frac{x_{Tol}}{x_{Iso}+x_{Hep}} - 0.514\right)^2 + P_{tol} \cdot \left(\frac{x_{Tol}}{x_{Iso}+x_{Hep}} - 0.514\right) + S_{iO} \cdot \left(\frac{x_{Iso}}{x_{Iso}+x_{Hep}} - \right.$$
$$\left. 0.223\right)^2 + P_{iO} \cdot \left(\frac{x_{Iso}}{x_{Iso}+x_{Hep}} - 0.223\right) + S_{etha} \cdot \left(\frac{x_{Eth}}{x_{Iso}+x_{Hep}+x_{Tol}} - 0.115\right)^2 + P_{etha} \cdot$$
$$\left(\frac{x_{Eth}}{x_{Iso}+x_{Hep}+x_{Tol}} - 0.115\right) + U$$

C_1					C_2				
Parameter	**Function coefficients**				**Parameter**	**Function coefficients**			
S_{lam} = poly2(p)	8.66E-6	-1.78E-3	8.85E-2		S_{lam} = poly2(p)	-2.25E-4	4.68E-2	-2.38E+0	
P_{lam} = poly2(p)	-1.31E-5	2.15E-3	-7.60E-2		P_{lam} = poly2(p)	3.59E-4	-5.87E-2	2.04E+0	
S_{EGR} = poly2(p)	2.84E-1	1.48E-7	-1.84E-5		S_{EGR} = poly2(p)	-6.52E-9	-3.18E-6	4.79E-4	
P_{EGR} = poly2(p)	-2.13E-7	3.00E-5	-5.33E-4		P_{EGR} = poly2(p)	6.15E-6	-8.70E-4	1.58E-2	
S_{tol} = poly2(p)	1.09E-6	-1.75E-4	6.45E-3		S_{tol} = poly2(p)	-3.03E-5	4.82E-03	-1.77E-1	
P_{tol} = poly2(p)	-2.20E-7	3.99E-5	-1.30E-3		P_{tol} = poly2(p)	5.05E-06	-9.23E-4	3.08E-2	
S_{iO} = poly2(p)	8.43E-6	-1.86E-3	1.09E-1		S_{iO} = poly2(p)	-2.17E-4	4.89E-2	-2.98E+0	
P_{iO} = poly2(p)	-1.27E-5	2.61E-3	-1.46E-1		P_{iO} = poly2(p)	3.31E-4	-6.88E-2	3.90E+0	
S_{etha} = poly2(p)	2.73E-5	-9.08E-3	5.50E-1		S_{etha} = poly2(p)	-6.43E-4	2.38E-1	-1.49E+1	
P_{etha} = poly2(p)	-5.79E-6	9.43E-4	-3.26E-2		P_{etha} = poly2(p)	1.67E-4	-2.72E-2	9.48E-1	
U = exp2(p)	5.4E-2	-2.3E-2	3.1E-3	8.1E-3	U = exp2(p)	-1.3E+0	-2.0E-2	-5.8E-2	9.7E-3

C_3		C_4	
Parameter	**Function coefficients**	**Parameter**	**Function coefficients**

Parameter				Parameter			
S_{lam} = poly2(p)	2.17E-3	-4.60E-1	2.38E+1	S_{lam} = poly2(p)	-9.25E-3	1.99E+0	-1.06E+2
P_{lam} = poly2(p)	-3.68E-3	5.97E-1	-2.04E+1	P_{lam} = poly2(p)	1.66E-2	-2.68E+0	9.07E+1
S_{EGR} = poly2(p)	1.21E-7	2.43E-5	-4.66E-3	S_{EGR} = poly2(p)	-7.61E-7	-7.64E-5	2.01E-2
P_{EGR} = poly2(p)	-6.57E-5	9.31E-3	-1.72E-1	P_{EGR} = poly2(p)	3.08E-4	-4.37E-2	8.12E-1
S_{tol} = poly2(p)	3.13E-4	-4.95E-2	1.81E+0	S_{tol} = poly2(p)	-1.43E-3	2.25E-1	-8.26E+0
P_{tol} = poly2(p)	-4.12E-5	7.65E-3	-2.76E-1	P_{tol} = poly2(p)	1.37E-4	-2.67E-2	1.38E+0
S_{iO} = poly2(p)	2.09E-3	-4.83E-1	3.04E+1	S_{iO} = poly2(p)	-8.97E-3	2.12E+0	-1.38E+2
P_{iO} = poly2(p)	-3.23E-3	6.78E-1	-3.91E+1	P_{iO} = poly2(p)	1.40E-2	-2.98E+0	1.74E+2
S_{etha} = poly2(p)	5.52E-3	-2.32E+0	1.50E+2	S_{etha} = poly2(p)	-2.03E-2	9.99E+0	-6.69E+2
P_{etha} = poly2(p)	-1.78E-3	2.90E-1	-1.01E+1	P_{etha} = poly2(p)	8.33E-3	-1.36E+0	4.78E+1
U = exp2(p)	1.2E+1	-1.8E-2	3.4E-1	1.2E-2	U = exp2(p)	-4.8E+1	-1.5E-2	-6.2E-1	1.4E-2

C_5

Parameter	Function coefficients		
S_{lam} = poly2(p)	1.46E-2	-3.21E+0	1.75E+2
P_{lam} = poly2(p)	-2.79E-2	4.47E+0	-1.53E+2
S_{EGR} = poly2(p)	1.60E-6	7.90E-5	-3.25E-2
P_{EGR} = poly2(p)	-5.35E-4	7.58E-2	-1.44E+0
S_{tol} = poly2(p)	2.42E-3	-3.78E-1	1.43E+1
P_{tol} = poly2(p)	-1.18E-4	2.59E-2	-3.85E+0
S_{iO} = poly2(p)	1.44E-2	-3.49E+0	2.34E+2
P_{iO} = poly2(p)	-2.27E-2	4.90E+0	-2.90E+2
S_{etha} = poly2(p)	2.65E-2	-1.60E+1	1.11E+3
P_{etha} = poly2(p)	-1.45E-2	2.36E+0	-8.59E+1
U = exp2(p)	7.7E+1	-1.2E-2	2.7E+0	8.0E-3

A4. Modeling the Influence of Injected Water on the Mixture Auto-Ignition Behavior

All model coefficients of the high-temperature ignition delay (Equations 4.13 and 4.14), the low-temperature ignition delay (Equations 4.15 and 4.16) and the temperature increase (Equations 4.17 and 4.18) models as well as the equations for calculating all model parameters can be found in the previous sections (A1, A2 and A3 respectively).

Here, the equations for the model parameters $A_{i,high}$, $B_{i,high}$, $A_{i,low}$, $B_{i,low}$, $C_{1..5}$ as a function of the boundary conditions and surrogate composition are expanded with coefficients for the influence of injected water, Table A.4. The subscript "old" denotes the model coefficients calculated as shown in Sections A1, A2 and A3 respectively. The water content is defined as percent of the fuel mass. In order to increase the model accuracy by capturing the partially significant changes in the influences of the boundary conditions with pressure, all model parameters were modeled as a function of the current pressure p.

Table A.4: High-temperature ignition delay, low-temperature ignition delay and temperature increase model extensions and coefficients for the influence of injected water.

$$\tau_{i,high} = A_{i,high} \cdot e^{\left(\frac{B_{i,high}}{T}\right)} \text{ with}$$

$$A_{i,high} = A_{i,high,old} \cdot exp\left(S_{wat} \cdot \left(\frac{m_{water}}{m_{fuel}} \cdot 100\right)^2 + P_{wat} \cdot \frac{m_{water}}{m_{fuel}} \cdot 100\right) \text{ and}$$

$$B_{i,high} = B_{i,high,old} + Q_{wat} \cdot \left(\frac{m_{water}}{m_{fuel}} \cdot 100\right)^2 + R_{wat} \cdot \frac{m_{water}}{m_{fuel}} \cdot 100$$

$$\tau_{i,low} = A_{i,low} \cdot e^{\left(\frac{B_{i,low}}{T}\right)} \text{ with}$$

$$A_{i,low} = A_{i,low,old} \cdot exp\left(S_{wat} \cdot \left(\frac{m_{water}}{m_{fuel}} \cdot 100\right)^2 + P_{wat} \cdot \frac{m_{water}}{m_{fuel}} \cdot 100\right) \text{ and}$$

$$B_{i,low} = B_{i,low,old} + Q_{wat} \cdot \left(\frac{m_{water}}{m_{fuel}} \cdot 100\right)^2 + R_{wat} \cdot \frac{m_{water}}{m_{fuel}} \cdot 100$$

$$T_{incr,fit} = C_1 \left(\frac{T_{low}}{100}\right)^4 + C_2 \left(\frac{T_{low}}{100}\right)^3 + C_3 \left(\frac{T_{low}}{100}\right)^2 + C_4 \left(\frac{T_{low}}{100}\right)^1 + C_5 \text{ with}$$

$$C_{1..5} = C_{1..5,old} + S_{wat} \cdot \left(\frac{m_{water}}{m_{fuel}} \cdot 100\right)^2 + P_{wat} \cdot \frac{m_{water}}{m_{fuel}} \cdot 100$$

Parameter	Function coefficients			Parameter	Function coefficients		
$\tau_{1,high}$							
$S_{wat} = poly2(p)$	-2.95E-9	493.2E-9	-17.02E-6	$Q_{wat} = poly2(p)$	1.78E-6	-295.7E-6	9.97E-3
$P_{wat} = poly2(p)$	1.07E-6	-193.3E-6	8.26E-3	$R_{wat} = poly2(p)$	-635.0E-6	113.3E-3	-4.53E+0
$\tau_{2,high}$							
$S_{wat} = poly2(p)$	-7.98E-9	1.33E-6	-47.00E-6	$Q_{wat} = poly2(p)$	5.97E-6	-984.0E-6	32.68E-3
$P_{wat} = poly2(p)$	1.95E-6	-337.8E-6	15.76E-3	$R_{wat} = poly2(p)$	-1.40E-3	236.0E-3	-8.42E+0
$\tau_{3,high}$							
$S_{wat} = poly2(p)$	-674E-12	337.4E-9	-24.61E-6	$Q_{wat} = poly2(p)$	335.2E-9	-304.0E-6	26.27E-3
$P_{wat} = poly2(p)$	578.7E-9	-87.39E-6	8.33E-3	$R_{wat} = poly2(p)$	-619.1E-6	89.11E-3	-8.21E+0
$\tau_{1,low}$							
$S_{wat} = poly2(p)$	417E-12	-35.29E-9	-1.07E-6	$Q_{wat} = poly2(p)$	-247.4E-9	21.48E-6	442.5E-6
$P_{wat} = poly2(p)$	131.5E-9	-32.99E-6	3.80E-3	$R_{wat} = poly2(p)$	-63.83E-6	16.86E-3	-1.86E+0
$\tau_{2,low}$							
$S_{wat} = poly2(p)$	864E-12	-55.81E-9	-4.95E-6	$Q_{wat} = poly2(p)$	-895.2E-9	76.31E-6	2.53E-3
$P_{wat} = poly2(p)$	201.5E-9	-75.75E-6	12.06E-3	$R_{wat} = poly2(p)$	1.08E-6	27.69E-3	-7.36E+0
C_1							
$S_{wat} = poly2(p)$	-17E-12	4.75E-9	-348.3E-9	$P_{wat} = poly2(p)$	-9.48E-9	1.76E-6	-81.08E-6
C_2							
$S_{wat} = poly2(p)$	316E-12	-104E-9	8.61E-6	$P_{wat} = poly2(p)$	281.6E-9	-52.25E-6	2.39E-3
C_3							
$S_{wat} = poly2(p)$	-1.61E-9	804.3E-9	-79.23E-6	$P_{wat} = poly2(p)$	-3.07E-6	567.4E-6	-25.90E-3
C_4							
$S_{wat} = poly2(p)$	-11.7E-12	-2.49E-6	321E-6	$P_{wat} = poly2(p)$	14.73E-6	-2.69E-3	123E-3
C_5							
$S_{wat} = poly2(p)$	12.61E-9	2.27E-6	-479E-6	$P_{wat} = poly2(p)$	-26.20E-6	4.74E-3	-219.5E-3

Printed in the United States
By Bookmasters